FIRES IN THE SKY

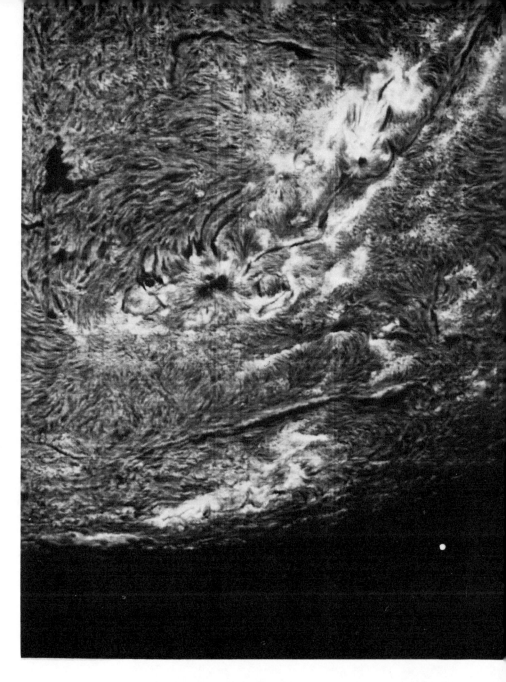

Like other stars, the Sun's surface is a churning sea of intensely hot gases, here photographed in the red light of hydrogen. The white dot at lower right represents Earth's size compared with the Sun. HALE OBSERVATORIES

Opposite: Detail from the *Rosarium Philosophorum*, 1550. In ancient times, eclipses of the Sun were believed to be caused by dragons and other such mythical animals taking bites out of the Sun.

FIRES IN THE SKY

The Birth and Death of Stars

BY ROY A. GALLANT

Four Winds Press   New York

LIBRARY OF CONGRESS CATALOGING IN PUBLICATION DATA

Gallant, Roy A
  Fires in the sky.

  Includes index.
  SUMMARY: Discusses the characteristics of stars
using the sun as an example, examining the composition
of the sun, various theories about its energy production, differences among stars, and the birth and death
of stars.
  1. Sun—Juvenile literature. 2. Stars—Evolution
—Juvenile literature [1. Sun. 2. Stars]
I. Title.
QB521.5.G34   523.8   78–4339
ISBN 0–590–07475–X

Published by Four Winds Press
A division of Scholastic Magazines, Inc., New York, N.Y.
Copyright © 1978 by Roy A.Gallant
All rights reserved
Printed in the United States of America
Library of Congress Catalog Card Number: 78–4339
1 2 3 4 5 82 81 80 79 78

For Tony, Linda, Richard, Paul, Sharon . . . and all you other stargazers by the dozens. May your luminosities never diminish.

ACKNOWLEDGMENTS

Several sections of this book are based on materials originally prepared for the University of Illinois Astronomy Program, on which the author served as a Science Education Specialist. These original materials are copyrighted © by the Board of Trustees, University of Illinois, and are used here by permission, although publication of the materials is not endorsed by the copyright holder. All sections of this book are based on a popular course in astronomy given for young adults and taught by the author at the American Museum-Hayden Planetarium, New York City.

My thanks to Doubleday & Company, Inc., for permission to use a brief passage ("Atomic Clocks") from *Discovering Rocks and Minerals,* by Roy A. Gallant and Christopher J. Schuberth, copyright © 1964, by Doubleday & Company, Inc.

My special thanks to Dr. Mark R. Chartrand III, Chairman of the American Museum-Hayden Planetarium, for reading and commenting on this book at the manuscript stage.

All diagrams and photographs, unless otherwise credited, are by Science Photo/Graphics, Inc.

CONTENTS

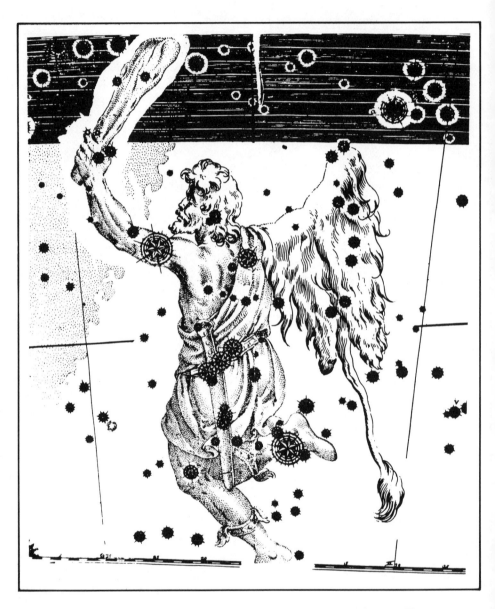

Orion, the Hunter.

PREFACE

Follow me among the stars

. . . to Rigel, the intensely bright star forming the left foot of the mighty sky hunter, Orion. It is a star that burns 60,000 times more brightly than the Sun. If Rigel were our local star, Earth's oceans would boil away and our planet would vaporize.

. . . to a dark cloud in the constellation Monoceros, which lies beside Orion in the winter sky. Here we can see dense, dark globes of gas and dust, stars just now being born.

. . . to a giant star in the constellation Auriga, just north of Orion. This star has a diameter 5,000 times greater than the Sun's. It is so large that if it were our local star, it would swallow up the Solar System out beyond the planet Saturn.

. . . to the Crab Nebula, in the constellation Taurus, which lies just northwest of Orion. Here are the remains of a star that was seen to blow itself to bits in the year A.D. 1054. It was one of those rare stars called supernovae, which astronomers are just beginning to understand.

. . . to a black hole in the sky, among the most puzzling objects in the Universe. The matter forming these mystery objects is packed together so tightly that a sugar-cube-sized piece weighs 100 billion tons.

These are some of the wonders you will find as we drift among the stars and try to discover how they are born, how they age, and how eventually they must die.

FIRES IN THE SKY

1.

FIRES IN THE SKY

From Astrology to Astronomy

If someone asked you what a star is, how would you answer? A pinpoint of light in the sky? Or a ball of glowing gas? Or an object like the Sun? Or possibly a lump of matter that shines with its own light?

All those answers are right, as far as they go. Now what if someone asked you how big an average star like the Sun is. How much matter does it have? What is it made of? What makes it shine? These questions are a bit harder to answer. But each one we manage to answer brings us closer to what a star really is, what is does, and how it does what it does.

It wasn't until the 1930s that astronomers began to discover what goes on in the deep core of a star like the Sun to make the star pour out huge amounts of energy. That isn't very long ago when you consider that astronomy as a science has been around for about 5,000 years. Around that time, and earlier, men were studying the stars not only to find out what they were and how they seemed to move, but for other reasons as well. They believed that the stars and planets were inhabited by spirits who held certain powers over kings, queens, and their subjects.

Today, most people no longer hold such beliefs, since the beliefs have no scientific basis. Yet many people look to the planets and stars for guidance in performing, or not performing, certain acts—a "lucky" or "unlucky" time to be married, to make a large financial investment, to sell property, to seek medical treatment, to have a child, and so on. Even such things as earthquakes, floods, and drought were mistakenly linked with influence of the stars. *Astrology*, as this belief is called, is nothing more than superstition and should not be confused with astronomy. The real importance of astrology is that eventually it gave rise to the science of *astronomy*, which is the study of all celestial bodies, their distances, brightness, sizes, motions, relative positions, what they are made of, and how they are put together.

Once the old sky-watchers turned their interest from supernatural spirits inhabiting the stars to a study of the motions of the stars and what the stars are made of, they were well on their way to becoming astronomers. Surely, ancient peoples looked on the Sun as the most "important" object in the heavens. It turned night into day and day into night as it appeared to trace an arc across the sky. In addition to providing light, the Sun also warmed the land; and as it climbed higher and higher above the horizon day after day, it warmed the land enough so that each year at the same time a reawakening of the world—or spring—occurred. It is no wonder that primitive peoples, even to this day, look on the Sun as a powerful god.

What is the Sun? Does it actually move across the sky as our eyes tell us it does? And is there a relationship between the Sun and the other stars? What happens to the stars at sunrise? Do they go out and disappear when the Sun turns night into day, and then just as mysteriously reappear out of nothingness after the Sun sets? By asking such questions and making observations to find answers, people living thousands of years ago were laying the foundation for not only astronomy, but science itself.

What Are the Stars?

These people had at least three ways of discovering that the stars remain in the sky always and exist quite independently of the Sun: (1) During a total solar eclipse, when the Moon blocks out the Sun, the stars gradually appear as more and more of the Sun's disk is hidden and the sky grows darker. (2) At dawn, just before sunrise, stars high overhead and those brighter stars down near the horizon stay visible for a short time while the dimmer stars fade from view. So even though the Sun can be seen peeping above the horizon, at least the brighter stars remain visible. (3) With the changing seasons, astronomers long ago must have noticed that certain groupings of stars—*constellations*—are visible during certain times of the year but not at other times. The observation that year after year the constellations repeat the same patterns of motion across the

Interest in what the stars are and the patterns of their apparent
motions as seen from Earth is as old as mankind. Shown here
are the remains of Stonehenge, an astronomical observatory built
in England over a period of 300 years, beginning sometime
around 2000 B.C. or earlier.

CBS TELEVISION NETWORK

sky would suggest that the stars do not dissolve and disappear by day and haphazardly reappear by night. Instead, the stars can be shown to be every bit as permanent and regular in their motion as the Sun is.

Such observations can be made by any of us, and without the aid of astronomical instruments. So we can safely say that people living several thousands of years ago were quite familiar with the Sun's yearly path among the constellations. Twelve constellations, called the *Zodiac*, make up that path that circles the entire sky, called the *ecliptic*.

The changing pattern of the stars that could be seen season by season must have been one of the first astronomical puzzles astronomers of ancient times tried to solve. About 5,000 years ago, astronomers in India imagined

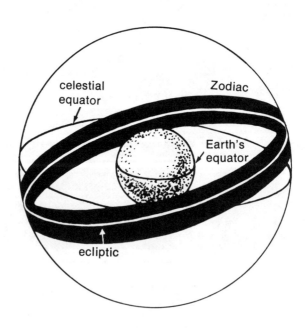

Earth as a disk-shaped object suspended in space. Hanging above it was a huge dome. By night and by day, the Sun, planets, and stars supposedly glided across the surface of this celestial dome.

Several centuries later, around 500 B.C., scholars of ancient Greece also offered explanations of what they could observe of the universe of stars and Earth's place in it. One, a teacher named Anaximenes, pictured Earth as a flat tabletop with moisture rising from it here and there. This moisture, he said, formed the stars. And since the moisture-substance rose, there could not be any stars beneath the Earth-table. The stars, he explained, move around Earth "as a cap turns round our head." He explained the Sun's disappearance at night by saying that it "is hidden from sight, not because it goes under Earth, but because it is hidden by mountains, and because its distance from us becomes greater."

Heraclitus, another Greek scholar, said that the stars, including the Sun, were made of fire and that each rested in a bowl. At night when we can see the stars, he explained, the mouths of the bowls are turned toward us and we see the stars' fire. But during the day, the bowls turn upward and hide the fire from view. Because the Sun-bowl was closest to Earth, he said, we received more light and heat from it than we do from the more distant bowls.

Still another of the ancient Greek scholars, Empedocles, pictured the Universe as a great, slowly turning shell. Located inside this shell, he said, Earth was in daylight when the fiery half of the shell was on top. But as the shell gradually turned, the fiery half slowly disappeared beneath and was replaced by the dark half. The Sun, he said, was not an object at all, but only a reflection of the fire source by the sky-shell.

Some supposed that the stars were glowing balls of molten iron. This idea isn't so farfetched, particularly if you have ever seen a meteor shower. *Meteors* are those fiery trails in the sky left by falling *meteoroids*—lumps of iron and stone—that enter Earth's atmosphere at high speed and burn up. When a meteoroid survives its hot journey through the atmosphere and reaches the ground, it is called a *meteorite*. To the Greeks of old it was perfectly reasonable to suppose that the hot, iron meteorities they found now and then were the remains of stars that had fallen to Earth.

Scientific Models

Such ideas about the stars may seem childish to us now, but these people did not know the answers and were searching for explanations based on reason rather than superstition and the supernatural. This was at a time long before the sciences of chemistry and physics had come into being. The ancient Greeks had no concept of the chemical elements, so they were extremely limited in the ways they could think about matter. And they had no formal understanding of the universal law of gravitation. Without an understanding of this law, they could not progress very far in an understanding of the many motions they could observe in the sky.

As long as this or that explanation of the universe of stars seemed to be in keeping with what could be observed, the explanation worked. We call such explanations *scientific models*. And astronomers, like all other scientists, are

"The stars fell like flakes of snow," said one observer of the
Leonid Shower of meteors, November 13, 1833. Another observer,
also believing that meteors were stars, thought that there would be
no stars left in the sky the next night.

THE AMERICAN MUSEUM-HAYDEN PLANETARIUM

forever making up scientific models to account for everything they can detect and measure. Our view of an atom is
just such a model, as is our view of DNA and other biological molecules. The history of the way scientists have viewed
atoms is a series of models, each later one including certain
changes based on new observations of how atoms behave
and of ever newer kinds of particles they contain.

So we mustn't be too hard on the old Greeks. They were
the first to reason correctly that Earth moves around the
Sun, although the idea was not generally accepted by scholars for more than 1,500 years! And they were the first to
show that Earth was a sphere, not a disk. They also were
the first to measure accurately Earth's size, and they correctly reasoned that the cloudy band of light we call the
Milky Way was made up of the countless stars.

It may seem that we know nearly everything we want to
know about the stars—how they move through space, how
they are born, how they age, and how they eventually go
out. But nearly any astronomer would tell you how very
little is known about the stars, simply because astronomers
realize there is so much more to be learned. This is what
scientists mean when they say that science is a never-ending
search. For each discovery is a window revealing new problems to be solved.

Today we are in a position similar to that of the old
Greeks. We can observe many things about the heavens that
puzzle us, and we build scientific models in an attempt to
explain those puzzles. The mysterious objects called "black
holes" are an example. And so is the astronomical puzzle
posed by a giant star that blows itself to bits as a

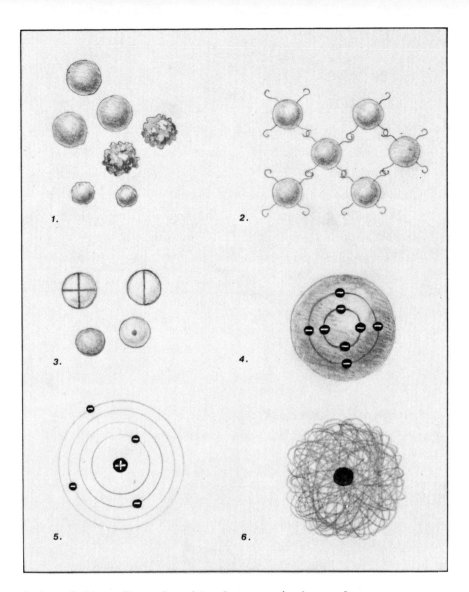

1. Around 450 B.C., Democritus pictured atoms as tiny lumps of matter that cannot be divided. Some were smooth, others rough, some large, some small. 2. Around 1630, Pierre Gassendi said that atoms must be hooked together in some way in order to form solids. 3. Around 1802, John Dalton said that atoms of different substances must differ in size and weight and that they combine as compounds. 4. Around 1897, J. J. Thomson pictured the atom as a positively charged globe with electrons inside. 5. Around 1913, Niels Bohr pictured the atom as having a positive nucleus surrounded by electrons orbiting in fixed shells. 6. Today we picture the atom as a cluster of positively charged protons and electrically neutral neutrons forming the massive nucleus, with electrons occupying certain fixed energy levels, the electrons being free to move from one energy level to another.

supernova. We have answers in the form of models to explain what we can observe. While some models may be on the right track but very incomplete, others are bound to turn out to be as far from the truth as was Heraclitus's model of the stars as bowls of fire.

We could spend many more chapters visiting with the old Greeks, and learning much from them. But instead, we will move the clock ahead more than 2,000 years to the time when astronomers and other scientists began to develop some of our "modern" ideas about the stars.

2.

THE SUN:
THE STAR WE KNOW BEST

How Big Is the Sun?

At least since the time of the ancient Greeks, astronomers have looked on the Sun as a star. It appears very much brighter to us than the other stars, they supposed, because it is so close to us. Not all astronomers agreed with this idea. Some supposed that the Sun was a special object associated with Earth. And Earth, they argued, was at the very center of the Universe. They regarded the stars as relatively small fiery objects all lying the same distance from Earth, each one attached to the inner surface of a great crystal sphere. Because the sphere slowly turned on its axis, the stars as a group were seen first to glide up over our heads, then slip down below the western horizon, and then up again over the eastern horizon night after night.

Everything that astronomers have been able to learn about the Sun and distant stars over the many centuries since ancient Greek times supports the idea that the Sun is nothing more than a nearby star. If so, then the best way to begin to find out about the stars is to study the Sun, since it is so close to us. And that's what astronomers began to do back in the early 1600s, when the Italian astronomer Galileo became the first to use a telescope to study the Sun, planets, and distant stars.

One of the first things astronomers wanted to know about stars is their size. So, how big is the Sun? If you held a measuring tape out at arm's length and measured the size of the windshield of a car parked way down the street, you'd come up with a measurement of four or five centimeters (cm) or two inches across. (An explanation of the metric system of measurement appears on page 145. It is used throughout this book, although equivalent units in the English system are also given.) Now you know from experience that that measurement can't be right. So you walk closer to the car, say half the distance you were before, and measure again. The size you measure this time is a little larger, but it still is not right. What you have been measuring so far is the *apparent* size of the windshield, or what it appears to be to your eye, depending on how far away you happen to be standing. To find the actual size, you would walk up to the car and hold the measuring tape across the windshield. Then you would be able to say, ''This windshield is 135 centimeters (54 inches) across.''

If you held your measuring tape up to measure the Sun's *diameter*, or the distance across its widest part, you'd come up with a measurement of one or two centimeters (about an inch). WARNING: *Never stare directly at the Sun. Its radiation can permanently damage your eyes.* Your

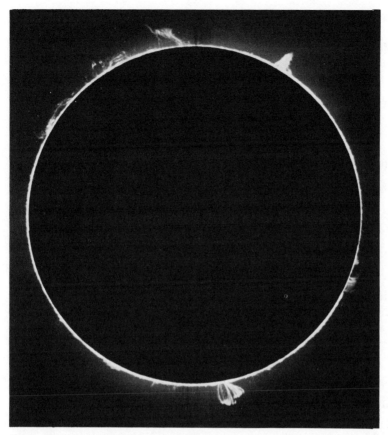

Like other stars, the "fiery" Sun is active. In this photograph
most of the Sun's disk has been blocked out to reveal several
eruptive surface features called prominences. HALE OBSERVATORIES

experience in measuring the car's windshield tells you that
the Sun's diameter has to be very much more than two centimeters. Since you can't walk right up to the Sun and measure it, you have to find some other way. If you know
how far away the Sun is, you can then use its apparent diameter—the one you measured—to find its actual diameter.
But how can we measure the Sun's distance?

Measuring the Sun's Distance

The first person known to have measured the Sun's distance from Earth was an ancient Greek scholar named Aristarchus, who lived during the third century B.C. Aristarchus used the Moon's motion around Earth as his "meter stick." What he did was estimate the angle between Moon Position 1, and Moon Position 2, as seen in the diagram. At Position 1, the Moon was seen to be exactly half full. At Position 2, the Moon formed a right angle (angle of 90°, or one-quarter of a circle) with Earth and the Sun. What Aristarchus wanted to know was the size of Angle A. That would tell him the Sun's distance, since that angle would become larger if the Sun were closer, and smaller if the Sun were farther away. It turns out that the Angle A is so small that it is just about impossible to measure accurately. Nevertheless, Aristarchus tried, and he came up with a distance to the Sun of 1,150,000 kilometers (km) (713,000 miles). Although his estimate was far short of the Sun's actual distance, it was a start.

About a century later another Greek scholar, named Hipparchus, measured the Moon's passage through Earth's shadow during a lunar eclipse and used this as a meter stick. Because it was so hard to measure the exact time the Moon passed into and out of Earth's shadow, Hipparchus's estimate of the Sun's distance was also wrong, and by a large amount. Although he estimated that the Sun was only 15,000,000 km (9,000,000 miles) away, he did suggest that the Sun was very distant indeed.

Other astronomers have used still other ways of measur-

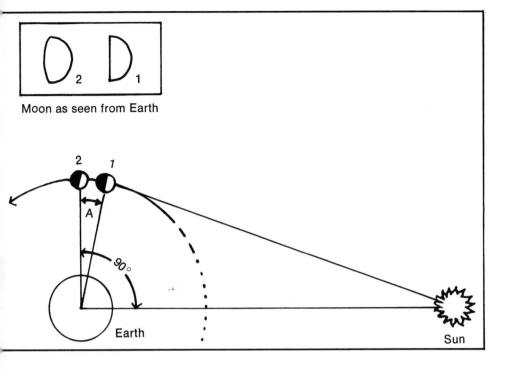

Moon as seen from Earth

Aristarchus used the Moon's motion around Earth as a method
of measuring Earth's distance from the Sun.

ing the Sun's distance. Over the years measurements have
become more and more accurate. Astronomers today use
such electronic tools as radar, artificial satellites, and rocket
craft put into orbit around the Sun. The United States
Naval Observatory in Washington, D.C., which is the head-
quarters in this country for all such measurements, presently puts the "official" distance of the Sun at 149,600,000
kilometers (92,752,000 miles). How accurate is that measurement? Various distances given by different electronic
devices differ by about 30,000,000 kilometers (18,600,000
miles). Astronomers often have to work with very large dis-

tances, and just as often it is difficult to make measurements as precisely as we would like. No astronomer would seriously argue that the Sun "actually" is a few hundred or even a few thousand kilometers more or less than 148,607,370 kilometers, which happens to be one precise distance obtained. For our purposes in this book, we can round off the Sun's average distance from Earth to 150 million kilometers (93 million miles). That distance is termed the *astronomical unit* and is often used as a convenient celestial measuring stick.

Now that we have a distance figure for the Sun, we can measure the Sun's diameter. And this is something you can do yourself. All you need is a shoebox, a piece of white paper, a piece of aluminum foil, a pin, and a ruler marked off in centimeters and millimeters (mm). (If you enjoy the challenge of solving simple math problems, you will want to do the following activity. But if math bores you, then skip on to page 24.)

Measuring the Sun's Diameter

Make a box like that shown in the diagram by cutting out a window in one end, taping a piece of aluminum foil over the window, and punching a small, *neat* hole in the foil with a pin. Next, tape a white paper "screen" to the inside far end of the box. On a sunny day prop the box up on a book or rock so that the foil window is pointed right at the Sun. If it is, you will notice an image of the Sun shining on the middle of the white paper screen. You can do this, by the way, either outdoors or indoors by a sunny window.

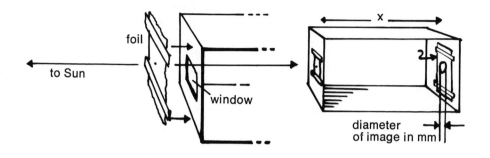

to Sun

foil

window

x

diameter
of image in mm

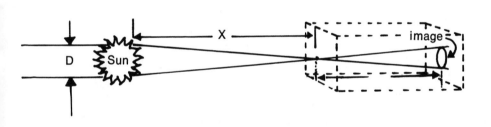

D Sun

X

image

D

X

x

d

If the Sun's image on the screen is too dim to measure, make the pinhole a *tiny* bit bigger, but be sure to keep the hole neat, meaning no ragged edges. When you have a clear image on the screen, very carefully reach into the box and with a fine ball-point pen or *sharp* pencil mark as accurately as you can each edge of the Sun's image. Now you can put the box in a more workable position and measure the distance, in millimeters, between the two marks. What did you get? (If you do not have a sharp image of the Sun, your measurement will be too large, which means that your figure for the diameter will be too large.) Next, measure the length of the box in millimeters. Since these first two measurements are in millimeters, you should also convert the Sun's distance of 150 million kilometers to millimeters, which comes to 150 trillion millimeters, as shown here:

$$150{,}000{,}000 \text{ km} \times 1{,}000$$
$$= 150{,}000{,}000{,}000 \text{ meters (m)} \times 100$$
$$= 15{,}000{,}000{,}000{,}000 \text{ cm} \times 10 = 150{,}000{,}000{,}000{,}000 \text{ mm}$$

Notice in the diagram that two triangles are formed by the Sun and shoebox setup. Since all three angles in the big triangle are exactly the same as the corresponding angles in the small triangle, we can use this equation to work out the Sun's diameter:

$$\frac{D}{X} = \frac{d}{x} \quad \text{or} \quad D = X \times \frac{d}{x}$$

D (the Sun's diameter) $= ?$
X (Sun's distance) $= 150{,}000{,}000{,}000{,}000$ mm
d (diameter of the Sun's image) $=$ your measurement
and x (length of the shoebox) $=$ your measurement.

When I worked this problem, I used an old box that some long envelopes came in. The length of the box was 313 millimeters, and the diameter of the Sun's image was 3 millimeters. If you work the problem with those figures, here is how it comes out:

$$D = \frac{150{,}000{,}000{,}000{,}000 \text{ mm} \times 3 \text{ mm}}{313 \text{ mm}}$$

$$= \frac{450{,}000{,}000{,}000{,}000 \text{ mm}}{313 \text{ mm}}$$

$$= 1{,}430{,}000{,}000{,}000 \text{ mm, or}$$
$$143{,}000{,}000{,}000 \text{ cm, or}$$
$$1{,}430{,}000{,}000 \text{ m, or}$$
$$1{,}430{,}000 \text{ km } (886{,}600 \text{ mi})$$

Both the answer above (1,430,000 kilometers) and the most precise measurements of the Sun's diameter that astronomers are able to come up with (1,392,000 kilometers) come to 1,400,000 kilometers when rounded off to two figures.

So now we know the size of one star among the several million visible through a small telescope. What about the sizes of all those other stars, and the billions more seen through a large telescope? How do we measure their sizes? We'll return to this problem in a later chapter. What we want to do next is find out how a star like the Sun shines.

3.

HOW DOES THE SUN SHINE?

Can the Sun Be Burning?

Knowing that the Sun's diameter is about 1,400,000 kilometers (868,000 miles) doesn't help very much when we ask how the Sun shines. Since the Sun gives off a lot of light and heat, which we can feel when we sun ourselves on a hot beach in summer, we might guess that the Sun is somehow associated with fire. Some scientists of ancient Greece thought that, too, but they imagined that the Sun's "fire" was a very special sort of substance unlike the fire from a burning candle or log.

The idea of a fiery Sun held on for many centuries. If the Sun is fire, then surely *something* must be burning to produce the fire. Whatever it is that is burning must be packed into a ball 1,400,000 kilometers in diameter, a ball that has been burning for as long as anyone can remember. From records going back 5,000 years, we can read about observations of the Sun. And there is nothing in those records to suggest that the Sun has been very much brighter or dimmer than it is now.

We can look back into the Sun's "burning" history even farther than 5,000 years ago. In 1952 biologists discovered in the Pacific Ocean 500-million-year-old fossil forms of a certain marine organism, known as *Neopilina*, still living today. If the Sun had been either very much brighter or very much dimmer then than it is now, these organisms probably would not have survived the change. They most likely would have become extinct. Since they did survive, this is evidence that nearly half a billion years ago the Sun was just about the same as we see it today. So if the Sun does give off its tremendous amount of light, heat, and other forms of energy by burning steadily, then it must be made of a very special substance indeed, a form of matter unknown to us on Earth. And that possibility, as you will find later, is very unlikely.

There is a rather simple way to test this idea of a burning Sun. Suppose that the Sun were a huge pile of wood, coal, or any other known substance that burns 1,400,000 kilometers in diameter. How long could it burn if it burned hot enough to give off as much heat and light as we observe the Sun to be giving off? It turns out that such a "burning" Sun could last only a few thousand years. That is far short of the hundreds of millions of years that we know the Sun has been shining brightly more or less in its present state. So the Sun cannot be an object that is burning chemically, as a log in your fireplace does.

It may surprise you to hear that the Sun is too hot to burn. On the scientific level, when we say that a piece of

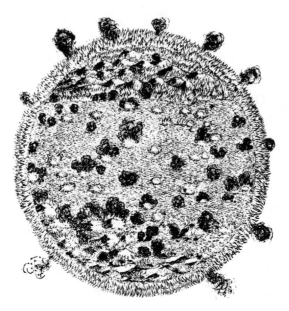

This drawing of the Sun made in the year 1635 suggests that it is
a fiery object with smokelike puffs being given off.

wood or paper is "burning," we mean that oxygen atoms of
the air combine with the substance burning. Paper, for
instance, is mostly cellulose, which is made up of carbon,
hydrogen, and oxygen atoms. When paper burns, its atoms
combine with atoms of oxygen, and the cellulose is changed
into carbon dioxide and water, with an ash of impure substance left.

To get something to burn, we have to speed up the motion
of its atoms by adding heat. One way of doing this is to rub
two sticks together. A lightning bolt striking a substance is
another way. Once the atoms of a substance are moving rapidly enough, they combine with oxygen and keep on doing

so until nothing is left of the substance, or until someone cuts off the oxygen supply by smothering the burning substance.

If you keep on speeding up the atoms or molecules of a substance—water, for instance—something interesting happens. The water molecules begin colliding with each other so vigorously that they fly off into the air. We then say that the substance has vaporized, or evaporated. When you boil a kettle of water, that's just what happens. Heat from the fire makes the water molecules bounce against each other so vigorously that the molecules of the top layer being struck by the speeding molecules below are knocked free and enter the air as the gas we call water vapor.

So the "hotter" a substance is—a solid, a liquid, or a gas—the faster its atoms or molecules are moving about. When we say that the temperature of the air, a piece of ice, or our bodies is thus and so, we are simply measuring the average speed of the atoms or molecules of the substance. For instance, when the molecules of air in a room are at what we call room temperature (68° F), their average speed is about 3 kilometers (2 miles) a second, which amounts to nearly 11,000 kilometers (6,700 miles) an hour.

By examining the color of the Sun's surface, astronomers can make a pretty good guess about its temperature. You know from experience that as a charcoal fire is allowed to go out, its coals change from white-hot to orange-hot, to red-hot as they cool off. And an object that glows bluish-white is hotter than a white-hot object. So as we look about us in the night sky and see stars of different colors, we know that the blue-white ones, such as Sirius and Rigel, are the hottest ones; the yellowish-white stars such as the Sun are medium-hot; and the reddish stars, such as Betelgeuse, are much cooler at their surfaces.

It turns out that the temperature of the Sun's surface is 6,000 K. (The convention is to drop the ° sign for K, but to include it for °F and °C.) So its atoms have to be moving at a

---

TEMPERATURE WITH A "K"

F, for Fahrenheit, is not the only temperature scale. Scientists use two others, which are used throughout this book.

Instead of having their thermometers marked in degrees F, most people of the world use a temperature scale marked in degrees C, for Celsius (also called centigrade). The C scale was invented by the Swedish astronomer Anders Celsius in the 1700s. As you can see in the diagram, the Celsius scale is the easiest one to use for everyday purposes. On it, ice melts at 0° and water boils at 100°, compared with the awkward numbers 32 and 212 on the F scale.

Objects, of course, can get much colder than freezing, or 0°C. But how much colder? Is there a limit? The British scientist William Thompson (also known as Lord Kelvin) thought that there was a limit. He supposed that if an object kept getting colder and colder, the atoms making it up gradually slow down and eventually stop moving. Atoms in that complete state of rest, he said, would have a temperature of *absolute zero*.

---

very high average speed. In fact, they are moving about so fast, and bumping into each other so hard, that they cannot combine. Instead, they simply bounce off each other and remain as single atoms. That is why we say that the Sun is "too hot to burn." Its atoms are moving too fast to link up with oxygen atoms and so take part in combustion.

The temperature scale based on Kelvin's idea of absolute zero is called the Kelvin scale (also the absolute scale). This scale is useful to measure how fast the atoms making up the air or other gases are darting about. The faster the atoms of a substance are moving about, the higher we say the object's temperature is. And this is true of the Kelvin, Celsius, and Fahrenheit scales alike.

Study the diagram to get used to the Kelvin scale. Notice that what we call "room temperature" is pretty close to 300 K. The boiling point of water is about 370 K. Also notice that on both the Kelvin and Celsius scales there are exactly 100 degrees between the freezing point and boiling point of water. On the Fahrenheit scale how many are there? This means that a Kelvin degree and a Celsius degree are larger than a Fahrenheit degree—nine-fifths larger. (If you want to find out what 86°F, for example, is on the C or K scale, or what 14°C is on the F or K scale, see the Glossary, which begins on page 131.

As we talk about the temperatures of stars, we will be using the Kelvin scale and will be dealing in temperatures of thousands and millions of degrees.

FAHRENHEIT	KELVIN	CELSIUS	
212°	373°	100°	water boils
98.6°	310°	37°	body temperature
70°	294°	21°	room temperature
32°	273°	0°	water freezes
—2°	254°	—19°	freon boils
—40°	233°	—40°	mercury freezes
—202°	143°	—130°	alcohol freezes
—460°	0°	—273°	absolute zero

A Shrinking Sun?

If the Sun and other stars don't give off energy by burning, then how do they shine?

Once scientists realized how far away the Sun is, it wasn't hard to realize that the Sun must give off tremendous amounts of energy. In one second the Sun radiates more energy than people have used ever since they have lived on Earth. During the 1850s, Kelvin and the German scientist Hermann von Helmholtz tried their hand at solving the solar-energy puzzle. Is it possible, they asked, that the Sun is slowly contracting, or shrinking, and that as it packs itself tighter and tighter around its core it gives off energy as a result? It seemed like an idea worth looking into. Again, science is a game of asking questions and then trying to find answers.

Kelvin and Helmholtz reasoned that an inward rain of Sun-matter falling toward the core would produce 10,000 times more energy than a burning Sun could. The in-falling solar particles would speed up the closer they approached the Sun's core, just as falling objects in Earth's atmosphere speed up the closer they approach Earth's center. Faster-falling particles in the Sun would mean higher temperatures. So a shrinking Sun seemed to account nicely for the observed energy output of our local star. The contraction hypothesis, as it came to be known, was favored for many years.

At the time Kelvin and Helmholtz lived, no one knew just how much energy the Sun radiates, so there was no way to test the hypothesis. But eventually a time came when it was possible to test it.

If the Sun and other stars give off energy by contraction, then they gradually must be shrinking. So a star like the Sun is ever changing in size, getting smaller and smaller year by year. To pour out as much energy as the Sun does, our local star would have to shrink by about 45 meters in diameter a year. While that is not very much in any one year, it adds up to quite a bit over the hundreds of millions of years the Sun is known to have been shining at its present rate.

Just how much does its diameter loss add up to? We computed the Sun's diameter at about 1,400,000 kilometers. If the Sun radiated energy by shrinking, and if its rate of shrinkage were 50 meters a year, then the Sun would decrease in diameter like this:

1 year	50 meters
10 years	500 meters
100 years	5,000 meters
1,000 years	50,000 meters
10,000 years	500,000 meters
30,000 years	1,500,000 meters
300,000 years	15,000,000 meters
3,000,000 years	150,000,000 meters
30,000,000 years	1,500,000,000 meters

or

1,500,000 km

So in a mere 30 million years a star the size of the Sun would lose its entire diameter if it were producing energy by contracting. According to this argument, 30 million years ago the Sun would have to have twice its present size. Since fossil evidence shows that the Sun has been shining at more and less its present rate for at least 400 million years, the contraction hypothesis must be thrown out.

As it turned out, it wasn't until the 1900s that astronomers worked out a new model accounting for how stars like the Sun shine, a model that fits all that we presently know to be true about the Sun. But before that model could be designed, astronomers first had to find out something about the Sun's *mass*, or how much matter it contains. Second, they had to learn about its *density*, or the amount of matter packed into a given amount of space.

How Massive Is the Sun?

Obviously it's impossible to weigh the Sun as we weigh a sack of flour or a pet canary. But there is a way, and it was invented by a bright young Englishman named Isaac Newton back in the 1600s. Studying to be a physicist and mathematician, Newton was home from college on a long vacation when he worked out his now famous universal law of gravitation, one of the most important scientific discoveries ever made.

We do not have to go into Newton's law of gravitation in very much detail to understand how it enables astronomers to calculate the masses of stars and planets, and to understand how it holds the Universe together. One principle of Newton's law is that any two objects in the Universe attract each other. Suppose that the Universe consisted only of two tennis balls. No matter how far away from each other they were, the tennis balls would attract each other, just as Earth and the Moon attract each other, and just as Earth and the Sun do. Newton also showed that the more massive two objects are, the greater the force of attraction between them. So two elephants in space would attract each other with more force than would two walnuts. Newton further showed that the closer two objects are to each other, the greater the force of attraction. And he was able to show mathematically exactly how gravitational force changes with distance and a change in mass.

Try to imagine the following situation, as shown in the diagram. It is gravitational attraction that keeps Earth in orbit about the Sun. And Earth's exact orbital path is dictated by the Sun's mass. In Part 2 of the diagram, notice that we have increased the Sun's mass. Also notice that because we have given the Sun more mass, Earth's orbit has changed. Because Earth and the Sun are now tugging at each other with more force than before, Earth's orbit is smaller and its orbital speed is greater. In Part 3 of the diagram, the Sun is less massive than in Part 1. Because the force of attraction is now less than it was originally, Earth has been allowed to move outward and settle down in a larger orbit, and has slowed down somewhat.

Stars, planets, or artificial satellites orbiting Earth all are affected the same way. For instance, notice in the table here how the orbital period and orbital velocity of an artificial satellite change as its distance from Earth increases.

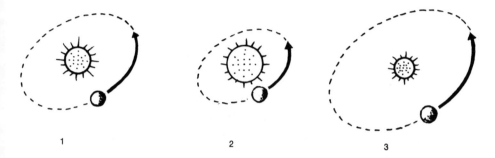

Earth's distance from the Sun and its orbital speed would
be different if the Sun's mass were either greater or less
than at present.

Period (in hours)	Distance From Earth (in kilometers)	Orbital Velocity (kilometers per hour)
2	1,688	25,319
7	12,207	16,676
12	20,240	13,933
17	27,195	12,406
21	32,273	11,562
24	35,871	11,059

With those facts in mind, we can now come back to the problem of just how much mass the Sun has. First of all, we know the size of Earth's orbit about the Sun because we know our distance from the Sun. Second, we know how long it takes Earth to circle the Sun. Since we know those two things, we can now use Newton's law of gravitation to figure out the Sun's mass. It is 2 grams with 33 zeros strung out after it, or in mathematical shorthand, 2×10^{33} grams. (See page 148 of the Appendix to find out how scientists deal with very large numbers.) Or, if you prefer big numbers, the Sun's mass is:

2,000,000,000,000,000,000,000,000,000,000,000 grams

or

4,405,000,000,000,000,000,000,000,000,000,000 pounds

or

2,200,000,000,000,000,000,000,000,000,000 metric tons

Compared with the Sun, Earth's mass is much less, only 6×10^{27} grams.

How Dense Is the Sun?

Before they could find out how the Sun shines, astronomers also had to know something about the Sun's density. *Density* is "mass per unit volume," or the amount of matter packed into a given amount of space. For example, if you had a box 1 meter along each side, the box would have a volume of $1 \times 1 \times 1$, or 1 cubic meter. Say that you packed the box full of Substance X without bulging out the sides or ends. Then you weigh the substance by pouring it out of the box onto a scale and find that it weighs 5.0 kilograms (kg). You can now say that the density of Substance X is 5.0 kilograms per cubic meter.

Density can be expressed in several different units of measure. For example, we can say that a substance is 15 pounds per cubic foot, or 8 ounces per cubic yard, 12 tons per cubic kilometer, and so on. But nearly always density is expressed in *grams per cubic centimeter*, which is written in shorthand as g/cc, or g/cm³.

How dense are you? If you think about it for a moment, it's pretty easy to figure out your density, if you know that water has a density of 1 g/cc (see table). If your density were very much greater than that of water, when swimming you would sink to the bottom like a lump of gold, which happens to have a very high density. And if your density were very much less than that of water, you would float easily, like a piece of cork. Since you can learn to float, with practice, then your density must be pretty close to that of water, or 1 g/cc. And so it is. Now compare the densities of these common substances:

SUBSTANCE	DENSITY	SUBSTANCE	DENSITY
Gold	19.3 g/cc	Your body	1.1 g/cc
Lead	11.3	Water	1.0
Iron	7.9	Cork	0.2
Earth (average)	5.5	Air (sea level)	0.0012

There is an easy way to find the density of any substance. Just weigh the amount of substance that is contained in a known volume, as we did with Substance X in the box measuring 1 cubic meter. Now what about the Sun? We already know how much it weighs, or its mass: 2×10^{33} grams. We also know the Sun's radius, which is one-half its diameter, which we measured. Knowing its radius, we use the equation below to find the Sun's volume, where V stands for volume and R for radius:

$$V = \frac{4\pi R^3}{3}$$

$$V = ?$$
$$\pi = 3.1$$
$$R = 7 \times 10^{10} \text{ cm}$$

Here's how you would solve the equation by doing some simple multiplication, division, and addition:

$$V = \frac{4 \times 3.1 \times (7.0 \times 10^{10}) \times (7.0 \times 10^{10}) \times (7.0 \times 10^{10}) \text{ cc}}{3}$$

$$= \frac{4 \times 3.1 \times 343 \times 10^{30} \text{ cc}}{3}$$

$$= \frac{4.2 \times 10^{33} \text{ cc}}{3}$$

$$= 1.4 \times 10^{33} \text{ cc}$$

So the volume of the Sun is 1.4×10^{33} cc, and its mass is 2.0×10^{33} grams. Now to find the Sun's density means solving one more equation, an easy one:

$$D = \frac{M}{V}$$

where D stands for density, M for mass, and V for volume. So,

$$D = \frac{2.0 \times \cancel{10^{38}} \, g}{1.4 \times \cancel{10^{38}} \, cc}$$

$$= 1.4 \ g/cc \ \text{(average density)}$$

A lot of work to find out such a small quantity, you might be thinking. But you'll find out how very important that "small" quantity is as we work our way into the Sun's interior and later begin examining blue giants, red giants, pulsars, and black holes. Without knowing something about the densities of those objects, astronomers would be hard put to find out what makes them work.

Before leaving the subject of the Sun's density, it is important to point out something about the figure 1.4 g/cc. It doesn't tell us anything about how dense any particular part of the Sun is! And that is exactly what we want to know. It tells us only that the *average* density of the Sun is 1.4 g/cc. In the same way, knowing that the average density of your body is 1.1 g/cc does not mean that is the density of your bones or of your hair or of your blood. The same situation is true for Earth, as you may have noticed in the table of densities. The *average* density of Earth is 5.5 g/cc. Rocks near Earth's surface have a density of only about 3.0 g/cc. But Earth's deep interior, made of a combination of iron and nickel, has a density of about 15.0 g/cc. If we knew Earth's average density only, we would not be able to say very much about how its density changes with depth toward the planet's core. In the same way, knowing the Sun's average density only does not tell us anything about how its density changes with depth toward the Sun's core.

So this is where matters stood until the early 1900s. Astronomers had managed to measure several properties of the Sun, including its distance, size, mass, volume, and average density. Further details were to begin coming thick and fast from the 1930s on. And our growing knowledge about the Sun as a star was to help astronomers answer the question posed by the ancient Greeks: "What makes the stars shine?" It also opened the way for an understanding of how all those other stars compared with the Sun.

4.

ENERGY FROM THE SUN

How Much?

How much energy does the Sun emit? Finding the answer to that question was an important step in finding out what makes the Sun and other stars shine.

Earlier, we said that in one second the Sun pours out more energy than people have used on this planet ever since there have been people. If you think about that statement for a moment, it helps make you realize how large an object the Sun is. All the energy given off by the Sun does not reach Earth, of course. Only the energy occupying that narrow path linking Earth to the Sun, a path the width of Earth's diameter, reaches us. All the rest simply goes off into space. It turns out that less than one-billionth of all of the Sun's energy reaches Earth.

Astronomers reasoned that if they could measure exactly how much energy the Sun emits, they would be in a much better position to figure out how that energy is generated.

As you were able to measure the Sun's diameter by using some simple apparatus, you can also measure the energy output of the Sun.

Measuring the Sun's Energy Output

MATERIALS: Here's what you'll need to build apparatus for measuring the Sun's energy output: a beaker marked in cc; a notebook for recording your data; a cardboard box about 30 cm long and about 4 cm deep; a baby bottle with nipple; a thin piece of cardboard large enough to cover the baby bottle; a thermometer marked in Celsius degrees.

WHAT TO DO: First, paint one flat surface of the baby bottle black and let it dry. Then fill the baby bottle with water, push the thermometer through the hole in the nipple, and screw the cap back on. (The nipple will be in its original position inside the bottle.) It is important that there is not a ridge of air in the bottle when you lay it on its side, so make sure the bottle is full. Now cut out a cardboard frame and tape it to the bottle *opposite* the painted side. When you cut this frame, try to keep the inside dimensions to the nearest complete centimeter; for example, 4 centimeters wide by 9 centimeters long, not 3.8 by 8.7 centimeters. This will simplify your arithmetic later. Now notch the cardboard box, as shown, so the thermometer can stick out through the notch opening. Now you're ready to set up the apparatus.

Wait for a warm, sunny day with clear skies. Around noon take the apparatus outdoors, put it in the shade, and watch the thermometer until it stops rising or falling. This may take up to five minutes or so. When the temperature reaches a point and stays there, record it in your notebook. Next take the apparatus out into the sunlight and prop it up, as shown, so that the Sun's rays are shining straight down onto the cardboard-framed surface of the bottle.

Using a watch with a sweep-second hand, record the temperature rise every 2 minutes for 10 minutes. What you are measuring is the amount of temperature rise during each 2-minute period. Your data might look something like this:

MINUTES	°C	TEMPERATURE CHANGE
0	27.5	(original reading in shade)
		. . . . .0.5
2	28.0	
		. . . . .0.3
4	28.3	
		. . . . .0.5
6	28.8	
		. . . . .0.5
8	29.3	
		. . . . .0.4
10	29.7	
		$\overline{2.2}$

2.2 ÷ 5 (readings) = 0.44°C *average* change each 2-minute period

After you have completed recording your temperature data, pour the water out of the baby bottle into the beaker and note how much water there was in cubic centimeters.

You will be calculating the Sun's energy output in calories. One *calorie* is the amount of energy it takes to raise the temperature of 1 gram of water 1°C. Look at the sample data again. Say that you measured 270 cubic centimeters of water in the baby bottle. That means that all the water in the bottle would have received a total of 0.44°C × 270 cc = 118 calories of energy every two minutes, or 59 calories per minute.

Now you're ready to work out the problem using your own data. First, figure out how many calories were received per minute by your own apparatus. Now you want to find out how many calories were received per square centimeter per minute. To find out, measure the inside dimensions of your cardboard frame. Say that they are 4 by 9 centimeters, giving an area of 36 square centimeters. To find out how many calories were received per square centimeter per minute, divide the total number of calories received per minute by the area of the frame, or 59 cal ÷ 36 sq cm = 1.6 cal/cm²/min.

Now suppose that you did this experiment indoors using a 200-watt light bulb in place of the Sun and that you positioned the light bulb 12 centimeters from the bottle. What do you think would happen if you then did the experiment a second time, but this time with the light bulb 24 centimeters away, then a third time with the light bulb 48 centimeters away, and a fourth time with the light bulb 96 centimeters away? You would probably guess, and correctly so, that fewer and fewer calories would be received per square centimeter of bottle surface each time. For the same reason, the taillights of an automobile fading off into the distance at night become less and less bright because less and less of their light reaches the area forming the pupils of your eyes. (If you're particularly adventurous, you might want to try this second activity to find out exactly what happens.)

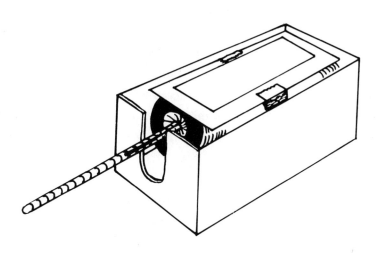

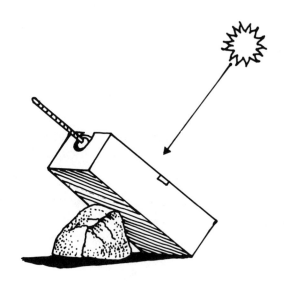

Astronomers all over the world have measured the amount of energy Earth receives from the Sun by using apparatus similar to that just described. And no matter whether they make the measurement at the Equator, Arctic Circle, or Antarctic Circle, they come up with the same results (after compensating for the thickness of Earth's atmosphere and certain other factors): Each square centimeter patch of Earth's surface receives two calories of energy from the Sun each minute. That is a very important measurement in astronomy, and it is called the *solar constant.*

Because the planet Mercury is closer to the Sun than Earth is, the solar constant for Mercury is greater than it is for Earth. And because Mars is farther away from the Sun than we are, the solar constant for Mars is less than it is for Earth. If Earth had a twin planet circling the Sun on the opposite edge of our orbit, the solar constant for that planet would be the same as it is for ours. In fact, any square-centimeter patch held anywhere in space at Earth's distance from the Sun would receive the same amount of energy— two calories per minute.

Finding the Sun's Luminosity

To find out the total amount of energy emitted by the Sun, we have to know how many square-centimeter patches there would be completely enclosing the Sun, each patch lying at Earth's distance from the Sun: 15×10^7 km. In other words, what we want to find out is the area of a sphere, every point on which lies at Earth's distance from

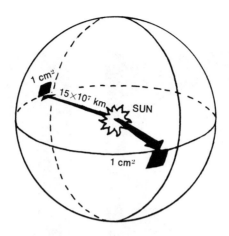

the Sun. Then, by multiplying the total number of square centimeters of surface of that sphere by the solar constant, we find out the Sun's total energy output, or *luminosity*.

That is an easy computation, since the area of a sphere is given by the short equation:

$$A = 4\pi R^2$$

in which A = area; π = 3.1; and R = the radius of the sphere, or Earth's distance from the Sun. Since we are working the problem in centimeters rather than in kilometers, we change 15×10^7 km to 15×10^{12} cm. So we solve the equation like this:

$$
\begin{aligned}
A &= 4 \times 3.1 \times (15 \times 10^{12} \text{ cm}) \times (15 \times 10^{12} \text{ cm}) \\
&= 12.4 \times (15 \times 10^{12} \text{ cm}) \times (15 \times 10^{12} \text{ cm}) \\
&= 186 \times 15 \times 10^{24} \text{ cm}^2 \\
&= 2{,}790 \times 10^{24} \text{ cm}^2 \\
&= 28 \times 10^{26} \text{ cm}^2
\end{aligned}
$$

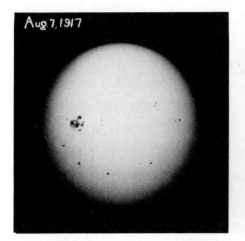
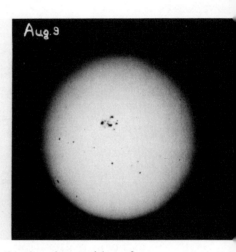

This sequence of photographs shows the changing position of a
group of sunspots over a six-day period, revealing the rate of the
Sun's rotation. YERKES OBSERVATORY

So there are 28×10^{26} square-centimeter patches making
up a sphere enclosing the Sun at Earth's distance. Since
each one of those square-centimeter patches receives 2 calories of energy each minute, then all we have to do is multiply
the sphere's surface area by 2 to find the Sun's luminosity:

$$\text{Luminosity} = 2 \text{ cal/min} \times 28 \times 10^{26} \text{ cm}^2$$
$$= 56 \times 10^{26} \text{ cal/min}$$

We have just added one more very important piece to
our model of the Sun, the total amount of energy pouring
out of it in a given period of time. Knowing that figure lets
us come up with still another important one—the Sun's surface temperature. By using another scientific law, we can
compute the Sun's surface temperature at 5,800 K. As you

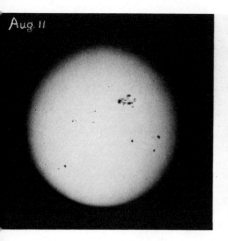

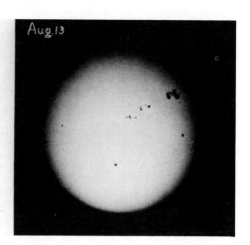

found earlier, the color of a star is also a clue to its surface temperature, so we have at least two ways of calculating a star's surface temperature.

Elements in the Sun

What more do we know about the Sun?

In the 1800s an instrument called the spectroscope was invented. Like a simple prism, a *spectroscope* separates white light into its different colors. By attaching spectroscopes to telescopes, astronomers can study and identify the makeup of the Sun's outer layers of gases. So far, about 70 of the 100 or more known elements have been found to make up the Sun's surface gases. The others probably are there but they fail to show up in the spectrum. But practically all

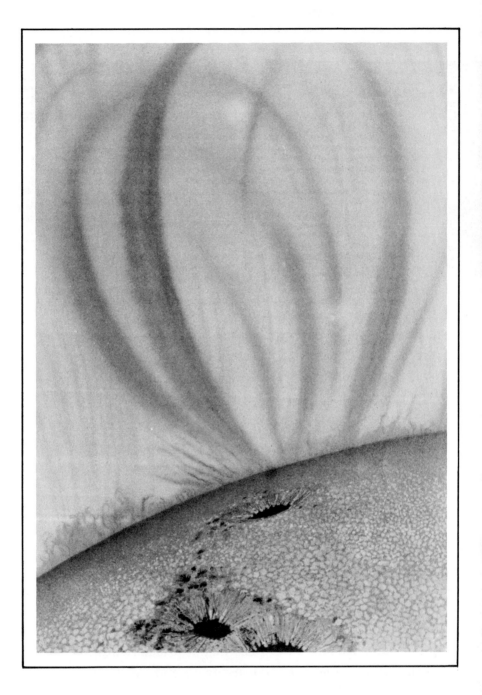

Artist's version of prominences erupting at the Sun's surface and looping upward to hundreds of thousands of kilometers. The black areas are sunspots. COURTESY ALDUS BOOKS LIMITED

the atoms of those gases turn out to be hydrogen and helium: 70 percent hydrogen and 26 percent helium. There are also some atoms of oxygen, fewer of nitrogen, then fewer of carbon, and then still fewer of neon.

In the 1600s, Galileo observed dark spots on the Sun. Since his time astronomers have learned much about sunspots. One bit of information they give us about the Sun is that it rotates on its axis once every 25 days. Other photographs of the Sun show it as an extremely active object. Gigantic explosive eruptions are taking place at its surface all the time. Some of these solar fireworks are shown on these pages. Eruptions of bright-appearing gas, called *prominences*, leap out to distances of 500,000 km (310,000 miles) and more. As astronomers observed more and more of these surface eruptions on the Sun, they began to think that the eruptions were caused by conditions deep within the Sun rather than at its relatively cool surface. So the more they could observe about the Sun's churning surface gases, the more questions they were able to ask about its mysterious hidden interior. And always one question was foremost in mind: What happens in the deep inner region of stars like the Sun to make them pour out so much energy, and to keep them shining at such a spendthrift rate for many hundreds of millions of years?

5

HOW GASES BEHAVE

Atoms in Motion

One of the things we can say about the Sun with certainty is that it is a large ball of intensely glowing gas. Its temperature is so high that its matter could not be held together by chemical means as we know solids, liquids, and certain gases to be held together on Earth.

To understand what the deep interior of the Sun, and other stars, must be like, it is necessary to know a few basic rules about the ways gases behave. Have you ever tried to force more water into a thermos bottle when the bottle was already filled right up to the top? Why can't you force more water in by hammering down on the cork? You know that you can keep pumping more and more air into a tire—until the tire bursts. So why can't you do the same with water?

Gases, but not liquids or solids, behave this way because gases are made up of a collection of atoms darting this way and that with "lots" of space between them. The atoms of a liquid are packed tightly together with practically no space between them, but they are loosely enough arranged so that they are free to slip and slide over each other. And that, of course, is why liquids flow. The atoms of solids tend to be locked into rigid patterns and vibrate in place rather than move over large distances.

Like liquids, gases flow, but they are much freer-flowing than liquids. Why does pumping more air into a tire cause the tire to get hard? Or, why does letting air flow out of a tire cause the tire to go soft? To understand exactly what is happening, we have to think of air, or any other gas, as a collection of atoms in motion.

Like a spoonful of pudding or a glass of water, a deep breath of air you take into your lungs has mass. It has mass because the individual atoms making it up have mass. Hydrogen, oxygen, nitrogen, helium, and the hundred or so other kinds of atoms known to us each has a certain mass, and each kind of atom has a mass different from the masses of all other kinds of atoms. That is because heavier atoms have more of the basic building blocks of matter than light atoms have. Hydrogen, for example, is the least massive atom of all, having a mass of 1 atomic mass unit, a unit of measure based on the mass of a carbon atom, whose mass is 12 atomic mass units. Helium is next after hydrogen with a mass of 4 atomic mass units. Oxygen has a mass of 15.9, lead a mass of 207.1, and uranium a mass of 238.0. So there are lightweight atoms and heavyweight atoms.

Now imagine two identical tires, both flat. You pump one full of hydrogen atoms. Then you pump up the other with exactly the same number of oxygen atoms. The oxygen tire will be heavier, or more massive, because its individual atoms have more mass than hydrogen atoms.

Try to imagine what happens inside a tire when you pump it up with any kind of gas. In come the atoms, pushed through the hose by the onrush of more atoms in the gas compressor outside. They enter the tire and bounce off each other and the inside surface of the tire and rim. Some of the atoms are moving slowly at a given moment, others rapidly. As a fast-moving atom collides with a slow-moving one, the slow-moving one is speeded up and bounces off in a new direction while the fast-moving one loses some of its speed. Throughout the tire, even after it is pumped up and the hose is detached, the direction and speed of each atom is purely a matter of chance. It all depends on what kind of collision an atom has just had.

If you could time the speed of each individual atom in the tire at a given instant and then divide all those speeds by the number of individual atoms, you would have the average speed of the atoms making up the gas. That average speed is a measure of the energy of motion—or *kinetic* energy—of the gas and can be expressed as a temperature. The atoms of a gas at room temperature have an average speed of about 3 kilometers per second, which can be expressed as a temperature of about 300 K. So, high gas temperatures mean high atomic speeds, and low temperatures mean low speeds. To know how a gas is behaving at a particular moment, it is important to know what the temperature of the gas happens to be at that moment.

If we next took our imaginary tire out of the shade and put it in the hot sunshine, what would happen to the gas

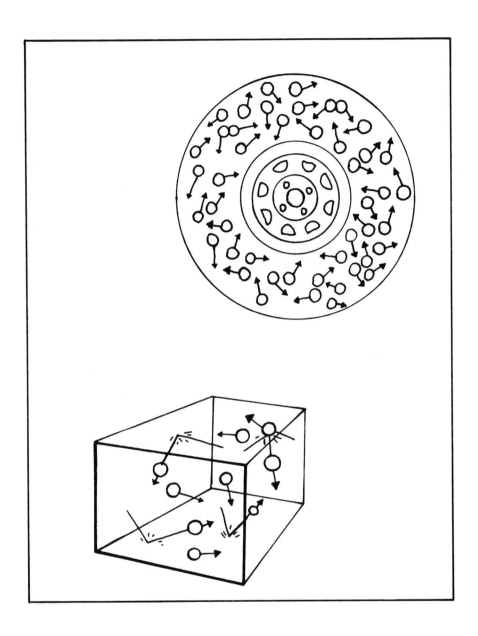

As in a tire or in a box, atoms of gas are continually moving about this way and that, colliding with each other and with the sides of their container.

inside? Energy is pouring out of the Sun and striking each square-centimeter patch of the tire at the rate of two calories per minute. That energy is going to increase the average speed of the gas inside the tire, or increase the temperature. As the individual atoms are speeded up, they begin to collide with great force, bouncing off each other and off the inside of the tire with greater force. The average force with which the atoms rebound from each other and the inside wall of the tire is called the *pressure* of the gas—or force per unit area. So while the temperature is a measure of the average speed of the atoms of a gas, the pressure is a measure of the average force of collisions.

It should be pretty obvious that if you increase the temperature of a gas, the pressure must also go up. This must be so because you are increasing the number *and* force of

At any given moment the number and force of collisions of atoms on one side of a partition will equal that on the other side. This is what we call the "pressure" of a gas.

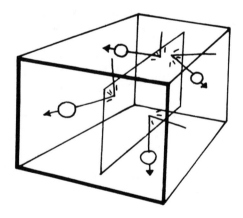

atomic collisions. And if you decrease the temperature, the pressure must go down because you are decreasing the number and force of collisions.

Can you think of a way of increasing the pressure of a tire without increasing the temperature? Simply by adding more gas atoms. In other words, by increasing the *number density* of the gas. While "density" is mass per unit volume, "number density" is simply the number of particles per unit volume, not considering mass. By adding more atoms we increase the number of collisions, because there are more atoms to have collisions. More atomic hits on the inside surface of the tire will make the tire feel harder from the outside. So we can raise the pressure of a gas by increasing either its temperature or its number density or, of course, doing both.

This model of a gas works well even though we can't actually see the atoms of a gas doing all of the things just described. From all that scientists are able to observe, the model works as well for a bicycle tire as it does for the air in a room or the gases spread thinly in space or those packed densely inside stars. The pressure of a gas anywhere in the Universe depends on two things—the temperature of the gas and the number density.

6.

ENERGY FROM
THE SUN'S CORE

Energy to "Burn"

Knowing how gases behave under different conditions of temperature, pressure, and number density, we can now add to our model of the Sun. Heat and other forms of energy are pouring out of the Sun at the rate of 56×10^{26} cal/min. Where is all that energy coming from? The Sun's surface gases cannot produce any energy, so the energy must be welling up from deep within the Sun.

Another reason for looking to the Sun's deep interior for the source of its energy is that heat can flow only in one direction—from a region of higher temperature to a region of lower temperature. You wouldn't expect heat to flow from the cool end of a poker to the red-hot end. The "cool end" of the Sun is its surface, with a temperature of 6,000 K. For the surface to remain at that temperature, heat and other energy must be welling up from within, the core region being the "hot end" of the Sun. And if the Sun's core region has a higher temperature than the surface, then the pressure of the core gases also must be greater than the pressure of the surface gases.

If that were not so, we would have a Sun that was collapsing in on itself. For our gas-ball Sun to remain its present size century after century, the pressure must increase toward the center.

Picture a boxcar volume of gas at the Sun's surface. Why doesn't it fall down into the core region? You don't fall toward the core region of Earth because Earth's surface is pressing up against you. But when you're swimming, you begin to fall toward Earth's center if you let all the air out of your lungs. Earth's gravitation tends to pull you down toward the center, toward the region of greatest mass.

The boxcar of solar gas at the Sun's surface doesn't have solid ground to hold it up. The reason it stays up and doesn't fall toward the solar core must be that there is another boxcar volume of gas of still greater pressure just beneath and holding up the surface boxcar of gas. And so at each deeper level within the Sun, there must be volumes of gas of increasing pressure right down to the very center, where the pressure is greatest of all.

So from what astronomers know about the way pressure and temperature must increase with depth into the solar interior, they can also calculate how the number density also must increase with depth. The *average* density of the Sun is 1.4 grams per cubic centimeter, but that does not say anything about the Sun's density at any particular depth. The density of the gases at the Sun's surface is much less than that average figure, far less dense than 0.0012 grams

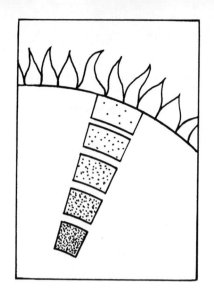

The Sun does not collapse in on itself because its various "layers" of gas are held up by denser layers of increasing pressure below.

per cubic centimeter, which is about the density of Earth's air at sea level. But in the core region the solar gases must be much more dense than that average figure, more than 160 grams per cubic centimeter, which is 15 times the density of lead.

So a model of the Sun that has temperature, density, number density, and pressure, all increasing toward the center, works. Although it accounts nicely for our observation of energy flowing out of the Sun, it does not account for the tremendous output of solar energy.

Mass into Energy

In 1905, Albert Einstein provided the clue that eventually led to the solution of the solar-energy mystery. It was his now famous equation that launched us into the Atomic Age:

$$E = mc^2$$

in which E stands for energy; m for mass; and c for the speed of light. In short, the equation says that mass and energy are interchangeable. Before his time, scientists had looked on energy and mass as being two very different things. Further, the equation shows that the enormous amount of energy locked up in a tiny bit of mass is staggering. For example, if you could convert a piece of chalk weighing only 4 grams entirely into energy, 84×10^{12} calories would be released. That's enough energy to raise the temperature of one million tons of water from just above freezing to 84°C!

Using Einstein's equation, two German scientists, Karl von Weizsäcker and Hans Bethe, in the 1930s hit on an idea that promised to explain how stars can pour out such vast amounts of energy each second. As so often happens in science, the two scientists came up with the idea quite independently. In brief, they said that small amounts of hydrogen in the hot and dense core region of the Sun are being changed into helium. As these changes take place, they said, tremendous amounts of energy are released, as Einstein's equation predicted. In other words, small amounts of mass (m) are being changed into large amounts of energy (E). So the Sun is shining by using its own mass as a source of fuel.

But how can hydrogen change into helium?

The Anatomy of an Atom

So far we have been talking about atoms as if they were solid lumps of matter like walnuts. As the diagrams here show, atoms are made up of three basic building blocks—electrons, protons, and neutrons. The simplest and lightest atom of all is hydrogen, which has one electron and one proton. An *electron* has one unit of negative electrical charge and so is represented by a minus sign (−). A *proton* has one unit of positive electrical charge and so is represented by a plus sign (+). As with the unlike poles of a magnet, an electron and proton attract each other. And as with the like poles of a magnet, two electrons tend to push each other away. The same is true of two protons. Since *neutrons* do not have any electrical charge, they neither attract nor repel each other, electrons, or protons. While electrons weigh practically nothing, protons and neutrons are much heavier by comparison and so make up just about all the mass of an atom. So, the bulk of the mass of any atom is in its central core, or *nucleus.*

The diagram shows that a hydrogen atom is electrically neutral. It does not carry a charge because the unit charge of the negative electron cancels out the unit charge of the positive proton. A helium atom also is neutral because its two electrons cancel out its two protons. Earlier we said that the mass of a hydrogen atom is 1 (its 1 proton) while the mass of a helium atom is 4 (2 protons + 2 neutrons). The atoms of other elements, such as oxygen, carbon, and uranium, have many more protons and neutrons forming the nucleus, and many more electrons. An atom of gold, for example, has 79 protons and 118 neutrons.

When we raise the temperature of a gas higher and higher, something interesting happens to its atoms. At about 300 K (room temperature), whole atoms are darting

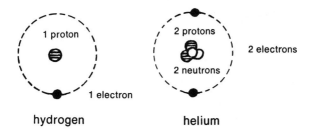

hydrogen   helium

about and bouncing off each other and the walls of their container. Raise the temperature to 10,000 K and the atoms collide so forcefully that their electrons are knocked away. What happens if the gas happens to be hydrogen? We end up with free electrons swimming about in a sea of free protons, all bumping into each other so hard that they just bounce off in a new direction at a new speed and for the most part stay detached from each other. But if we raise the temperature still more—to about 10,000,000 K—something very interesting happens to the protons: They are now darting about so fast and colliding so vigorously that their invisible energy bumpers no longer work, and two protons may fuse as a single lump of matter.

Hydrogen into Helium

Von Weizsäcker and Bethe pictured helium nuclei being produced in the core of the Sun by just such forceful collisions of hydrogen nuclei, numerous hydrogen nuclei coming together in a series of steps, fusing, and forming helium nuclei.

Although the process is much more complicated than shown here, this is pretty much what seems to happen. Two protons collide and fuse. As they fuse, one of the protons loses its electrical charge of $+1$ and turns into a neutron. During the change, some of the proton's mass is turned into energy, according to Einstein's equation, $E = mc^2$.

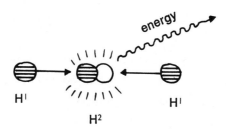

So now we are left with $H^1 + H^1 \longrightarrow H^2 +$ energy. (H^1 means a hydrogen nucleus with one proton; H^2 a hydrogen nucleus with two protons, etc.) Next we can picture one of the fused pair of particles colliding with another free proton, fusing with it, and releasing energy:

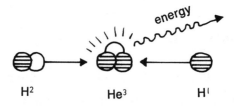

So now we have H² + H¹ ——→ H³ + energy. Next we can picture the trio of two protons fused with a neutron colliding with another such trio. During this collision two protons are freed and another small parcel of energy is emitted:

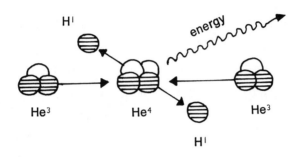

So now we have He³ + He³ ——→ He⁴ + H¹ + H¹ + energy. A nucleus of helium (He⁴) has been gradually built up by the collision of hydrogen nuclei (free protons) and added to the Sun's core. And during the process mass has been changed into small parcels of energy. Weizsäcker and Bethe pictured trillions of such fusions taking place each second within the Sun's core. And gradually the energy released works its way up through that enormous mass of overlying gases and breaks through the Sun's surface gases as light, heat, x rays, ultraviolet rays, and other forms of energy.

Do such fusions of hydrogen nuclei into helium nuclei actually take place, or was it only a theory? On the basis of that theory, it seemed possible to make a nuclear fusion device, one that would release a great outburst of energy. Such a device was indeed made and was proof that the fusion of hydrogen nuclei is possible—the hydrogen bomb. And scientists are now trying to develop nuclear fusion devices that release energy in a slower and controlled way. Although such controlled fusion devices can be started, the difficulty so far has been to keep them going. If scientists can solve that problem, the world's energy ills will be cured.

So nuclear fusions of hydrogen into helium can indeed take place. And the Sun's core, with its extremely high temperature of more than 15,000,000 K, its enormous density of more than 100 g/cc, and its crushing pressure, seemed to provide just the right conditions for this sustained explosive reaction to take place. But is that actually what causes the Sun and other stars like it to shine? Was it possible to test the idea, to find convincing evidence that it is so?

$$7.$$

LIFE-SPAN AND
AGE OF THE SUN

The Sun's Life-Span

One way of testing the idea that nuclear fusions of hydrogen into helium are the Sun's source of energy output was to figure out the life-span of the Sun. If the Sun keeps shining by changing mass into energy, then obviously the Sun loses some of its mass day by day, year by year, and century by century. Since the Sun is known to have been shining for hundreds of millions of years, then it must have lost quite a bit of mass through the fusion of hydrogen nuclei. And there must come a time when the hydrogen nuclei fuel tank of the Sun reads EMPTY. When will that be?

We can begin by asking how much of the Sun's mass is being changed into energy each minute. We know that the Sun gives off 56×10^{26} cal/min; and we know that when one gram of any kind of matter is changed completely into energy, it produces 21×10^{12} cal/g. So to find out how many grams of matter the Sun loses each minute, we divide the first figure by the second:

$$\frac{56 \times 10^{26} \text{ cal/min}}{21 \times 10^{12} \text{ cal/g}} = 2.7 \times 10^{14} \text{ g/min}$$

So, each minute the Sun is losing 2.7×10^{14} grams of mass, and it has been doing so ever since it has been shining as we know it to be shining today. That amounts to 3.9×10^{17} grams every day, and 1.4×10^{20} grams every year. At that rate of using up its mass, how much longer can the Sun keep shining?

That, of course, depends on how much hydrogen fuel it has left. Hydrogen nuclei can fuse only when the temperature is up around 10,000,000 K. The Sun's temperature is not that high throughout. The temperature keeps falling off farther and farther out from the core and is only about 6,000 K at the surface. Only in the core region of the Sun is the temperature high enough for the fusion of hydrogen to take place. So the core region is the Sun's hydrogen fuel tank, and that tank happens to hold about 10 percent of the Sun's total mass of 2×10^{33} grams. So 10 percent of that is 2×10^{32} grams.

As hydrogen nuclei are fusing and forming helium nuclei, parcels of energy are emitted. We get lots of energy at the expense of very little mass, as you saw in the example of raising a million tons of water from the freezing point to nearly the boiling point with only a small piece of chalk. So almost all the mass of hydrogen being fused into helium goes into the formation of new helium. Only a tiny amount (0.007) is changed into energy.

The Sun has a grand total of 2×10^{32} grams of hydrogen nuclei that can take part in nuclear fusions in the core region. What we want to find out now is what fraction of that grand total of mass is going to be lost as energy during the total life-span of the Sun. We find the answer by multiplying the grand total of mass available for fusions by the fraction lost during the fusion process:

$$2 \times 10^{32} \text{ grams} \times 0.007 = 1.4 \times 10^{30} \text{ grams}$$

So the total amount of mass lost by the Sun during its entire life-span comes out to be 1.4×10^{30} grams. Now we are in a position to make our final calculation to find out what the life-span of the Sun may be. To find that figure, we divide the total mass the Sun has in its hot core for hydrogen fusions by the amount of mass it keeps losing every year:

$$\text{Life-span} = \frac{1.4 \times 10^{30} \text{ grams (mass available for fusions)}}{1.4 \times 10^{20} \text{ grams (mass used per year)}}$$

$$= 1 \times 10^{10} \text{ years, or 10 billion years}$$

So the Sun appears to have a life-span of about 10 billion years, assuming that it will continue to emit energy in the future at about the same rate it has been emitting energy for the past several hundreds of millions of years. After 2,000 years of wondering what the stars are, how they shine, and how long their fires burn, astronomers have found some very satisfying answers, answers that did not come easily, or quickly.

How Old Is the Sun?

We have worked out the life-span of the Sun. But we are left with another question just as important. How old is the Sun? Knowing only that the Sun has a life-span of some 10 billion years tells us nothing about how much longer it has to keep shining. For example, if I told you that my dog Zubenelgenubi (named after the brightest star in the constellation Libra) has an expected life-span of about 12 years, you wouldn't know whether she was a puppy of 6 months or an adult dog 10 years old.

To find out how much longer the Sun can be expected to keep shining as the kind of star we know it to be today, we must find out how long it has been shining so far. In other words, how old is it?

We cannot reach up and tear a chunk out of the Sun to measure the age of its material. Therefore, we are forced to examine those bits and pieces of the Solar System that we can get our hands on. These include pieces of Earth rock, Moon rock, and cosmic stuff that rains down on Earth as meteorites. Scientists use "atomic clocks" to measure the ages of certain fossil remains and rocks found over various parts of Earth's surface. If you want to find out how these atomic clocks work, read the Appendix section beginning on page 149.

In brief, since the late 1800s, scientists have known that the atoms of certain elements said to be radioactive—such as uranium, potassium, and thorium—break down naturally. They break down by giving up bits and pieces of themselves. In the process, they are said to "decay" into the atoms of other elements. For example, uranium turns into lead, and so does thorium. Potassium turns into the element argon. These elements change from one kind into another at a known rate. So a rock containing uranium and its decay product lead can be dated by measuring how much decay product (lead) there is in relation to the parent product (uranium).

Geologists have used various radioactive elements to date rock samples from every corner of Earth and have come up with several ages. The oldest rocks from Earth's crust yet found are from Greenland and are 3.8 billion years old. Rocks from other regions—for example, near the Great Lakes and in certain parts of Europe—are 3.5 billion years old. But that does not mean that Earth was formed only 3.8 billion years ago. That age is for the crustal rock only, rock that went through its most recent geological change that long ago. While Earth has a recorded history going back at least 3.8 billion years, its history before that, or prehistory, is unknown to us by means of direct measurement.

The oldest Moon rocks that have yet been examined have a geologic history going back nearly a billion years earlier than the oldest Earth rocks. This makes sense if Earth and the Moon formed at the same time. Since the Moon is smaller than Earth, it would have cooled faster. Meteorites now appear to be the senior citizens of the Solar System. Radioactive dating of many stone and iron meteorites shows an average age of about 4.6 billion years. If we accept that as an approximate age of the Solar System, then Earth went through about 1 billion years of geological and chemical evolution before a crust of solid rock developed. So all the evidence points to an age somewhere around 5 billion years for the formation of the Sun.

If the Sun has been around for about 5 billion years, and if it has a life-span of about 10 billion years, then it is just about in middle age. So it has an extremely long life ahead of it considering the mere tick of the cosmic clock that humans have lived.

So we have finally hit on an energy source of the Sun, and one that is in keeping with all that we can observe about the Sun and its family of planets. A burning model for the Sun did not work for two reasons. First, the Sun is too hot to burn. And second, a burning Sun only 14×10^5 kilometers in diameter could last only a few thousand years. A contraction model of the Sun did not work because such a Sun could last a mere 30 million years. A hydrogen fusion model for the Sun does work because it can account for the very long time the Sun has been known to have been pouring out energy at the staggering rate of 56×10^{26} cal/min.

Journey to the Center of the Sun

So putting together all these bits and pieces of information about the Sun, we can paint this portrait of our solar furnace: The surface is a wild, erupting sea of gases, nearly all hydrogen but with smaller amounts of helium, oxygen, nitrogen, carbon, and other elements. The average temperature of these surface gases is about 6,000 K, and their pressure and density are less than that of the air at Earth's surface, a density of perhaps only 0.000,01 g/cc.

As we worked our way down through the Sun toward the core region, we would most likely find a more or less gradual increase in temperature, pressure, and density of nuclei.

At about one-fifth of the way to the center we would find temperatures close to 1,000,000 K and a density of about 0.01 g/cc. Our view probably would be a completely foggy one and our ears most likely would be bombarded with a continuous explosive rushing of wind, winds that would make the wildest hurricane on Earth seem like a balmy summer breeze. Halfway to the core, 95 percent of the Sun's mass would still lie ahead of us, only 5 percent being behind us. This is because the deeper regions of the Sun become increasingly dense and so contain most of the Sun's mass. At the four-fifths mark, nearly at the border of the fusion zone, the temperature would be about 8,000,000 K. And if it were possible to withstand the crushing pressure, it would be like being encased inside a solid steel cube with a density of about 40 g/cc. But our cube prison would not be dark. Instead it would be blindingly bright, lighted by the countless numbers of particles of light flooding outward from the zone of fusion within the solar core. Inside the fusion zone itself, the temperature would rise to at least 15,000,000 K, and the density of our solid gaseous prison would be more than 100 g/cc.

Not only is the Sun an inhospitable place to live, but an impossible place to visit. Exposed to its deadly radiation just outside the protective shield of Earth's atmosphere, in only a few minutes you would be thoroughly cooked. No object, no matter of what Earthly material it is made, can long survive as it is brought ever closer to the Sun. Gradually its atoms become so energized that they vibrate apart from each other and the object becomes vaporized. Yet, within Earth's protective atmosphere, plants and animals alike depend on the Sun's life-giving energy for growth and survival.

If the ancient Greeks and the ancient Babylonians before them were alive today and told of the awful powers of the Sun and of its long life, they probably would nod in agreement and say: "You see, we were right all along. The Sun is all-powerful and the giver of life. And that is why it surely must be ranked as a god."

8.

TWINKLE, TWINKLE, GIANT STAR

Luminosity: How Bright Is Bright?

You now should have some feeling for the Sun as a star. As you'll find throughout the rest of this book, our model of the Sun as a star will make what we have to say about other stars much more meaningful and interesting.

Like astronomers of old, perhaps you have studied the night sky and compared this or that star with others. While most of the stars appear very much alike, some catch our attention with some special feature or another. On the basis of such informal observing, we conclude that the stars are not all exactly alike.

One thing our eyes tell us is that stars come in a variety of colors—some pink, others red, or yellow, or bluish-white. This observation about the stars immediately tells you something important about them because the color of a star is an excellent clue to its surface temperature. The reddish stars are the coolest ones, the yellowish ones somewhat hotter, and the bluish-white stars the hottest.

Another thing you notice as you scan the night sky is that some stars appear brighter than others. But is that one which you see almost directly overhead, and which appears very bright, actually brighter than the dimmer-appearing one you see down near the eastern horizon? What would happen to the Sun's "brightness" if our local star suddenly left the Solar System behind and went speeding off toward the other end of the Galaxy?

We can't, of course, tell very much about the brightness of a star without knowing how far away it is, although there are some exceptions which we will deal with later. "Brightness" is a fuzzy word that doesn't tell us very much about one star when compared with another. Astronomers speak of the *apparent brightness* of the stars, which is how much of their light reaches our eyes. That is a far cry from the total amount of light and other forms of energy pouring out of that star in all directions of space. The apparent brightness of a star can be important, but astronomers also want to know a star's total energy output, or *luminosity*.

An example, which all of us have experienced, will make the difference between luminosity and apparent brightness clear. Suppose that one automobile is approaching another at night along a straight section of road, and that both have their headlights on high beam. At first, the lights of the approaching car seem to be a single fuzzy dot off on the horizon. But as the car comes nearer, the ball of light separates into two distinct points and the light from them appears brighter. And it continues to appear brighter right up until the car passes in the opposite direction. Did the lights themselves actually increase in luminosity, or did it just appear that way?

An astronomer would say that the luminosity of each headlight is unchanging and can be expressed in units called candlepower. He would refer to the changing "eye-brightness" of the moving lights as apparent brightness, which keeps changing with distance and which he could measure with an ordinary camera light meter.

The Stars: How Far Is "Far"?

Before astronomers could say very much about the luminosity of stars other than the Sun, they had to find out how far away this or that star was. It wasn't until 1838 that a distance to a star other than the Sun was discovered. The astronomer given credit for the discovery was the German Friedrich Bessel, and the star is the one listed in astronomical catalogs as 61 Cygni, a member of the constellation Cygnus the Swan. The method Bessel used is called the parallax method and can be used for stars lying nearly 500 light-years away.

A *light-year* is the distance light travels in one year at the rate of 299,000 kilometers (186,000 miles) per second, which comes to about 10 trillion kilometers (6 trillion miles). We can also have units such as a light-second, a light-minute, a light-day, and so on. For instance, the Moon is a little less than 2 light-seconds from Earth; Mars at its closest is about 4 light-minutes from Earth; the Sun, about 8 light-minutes; Pluto, about 5 light-hours; and the Sun's nearest stellar neighbor, Alpha Centauri, 4.3 light-years.

Measuring stellar distances by the *parallax method* is quite simple in principle. As the diagram shows, the astronomer first photographs the star whose distance he wants to know in January, say. Six months later, when Earth has completed one half of its orbit around the Sun, the astronomer photographs the star again. He then compares the two photographs, carefully measuring the amount by which the star appears to have shifted its position in relation to the more distant background stars. Some simple arithmetic then gives the distance to the star.

One way astronomers measure the distance of stars is the parallax method, but this method works only for stars that are relatively near.

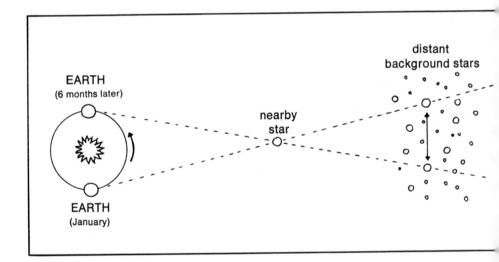

Look at the diagram on this page and notice that the star whose distance we want to know forms the center of a circle. Notice also that Earth's positions on opposite extremes of its orbit about the Sun also are shown, the distance on each side of the Sun being one astronomical unit (A.U.). If we can find the diameter or the radius of the circle around the star, we then know the distance of the star from the Sun. Once we know the distant star's parallax shift in degrees, the rest is easy. Say that the angle of parallax shift is 10°. What we want to find out next is how many astronomical units it is around the circle, and we can do that because we know that one A.U. traces out a distance of 5° (half of the parallax shift) on the circle circumference. Now a circle has 360°. So by dividing 5° into 360°, we find that the circle we are dealing with has a circumference of 72 A.U.

Knowing the circumference, we can now find the radius by the equation $C = 2\pi R$, where C is the circumference; $\pi = 3$; and R is the radius:

$$C = 2\pi R$$

$$R = \frac{C}{2\pi}$$

$$= \frac{72 \text{ A.U.}}{2 \times 3}$$

$$= \frac{72 \text{ A.U.}}{6}$$

$$= 12 \text{ A.U.}$$

So if the parallax shift of a star happened to be 10°, then the star's distance from us would be 12 A.U. But that is

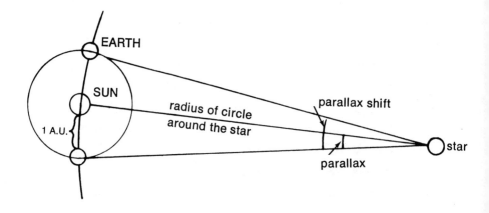

Finding the size of the parallax angle of a star is the first step in computing the star's distance.

impossible, of course, since 12 A.U. doesn't even get us out of the Solar System. Pluto, for instance, is 40 A.U. from the Sun. No, the parallax of stars is not measured in whole degrees, not even in minutes or seconds of degrees, but in small fractions of seconds of a degree. The parallax for Alpha Centauri, for example, is 0.76 second.

If you enjoy solving problems like this one, work out the distance to Alpha Centauri and see if it agrees with the distance given earlier. Also, work out the distance to a star with a parallax of 0.21 second, and another with a parallax of 0.04 second. In all cases these numbers are the parallax, or half the total parallax shift. Use them as we used the 5° (not 10°) value in the problem just solved.

To date, astonomers have measured the parallax for about 6,000 stars. The farther away a star is, the smaller its parallax shift is. The great majority of stars are so very far away that they do not show any parallax shift at all. So to find the distances to these other stars farther away than almost 500 light-years, astronomers had to think up another method of distance measurement.

A Law of Light

Astronomers can work out the distance to stars that are not parallax stars by using a law that describes what happens to a candle flame, flashlight, or any other light source as it moves away from us, or we from it. Let's return to our example of the apparent brightness of the headlights of a distant car approaching us on the highway at night. When the car has closed the distance from us to one-half its distance when we first saw it, its headlights now appear not twice as bright as before, but four times $(2 \times 2 = 4)$ as bright. When it has closed the distance to one-third, the headlights appear nine times $(3 \times 3 = 9)$ as bright. If we next watched its taillights fade off into the distance behind us, just the opposite would occur. Double the distance and the light source appears only one-fourth $(\frac{1}{2} \times \frac{1}{2} = \frac{1}{4})$ as bright as before. Triple the distance and it appears only one-ninth $(\frac{1}{3} \times \frac{1}{3} = 1/9)$ as bright. Because the arithmetic works out this way, this law is called the *inverse square law of light*. The force of gravitation strengthens and weakens the same way.

Knowing that light behaves in this way, the astronomer can attach a sensitive photocell, or fancy light meter, to a telescope and measure how bright a certain star appears to his photocell eye. If the astronomer knows the star's luminosity, or actual brightness, he can then compute the star's distance by applying the inverse square law of light.

This law of light gave astronomers an important way to measure the distance of stars farther away than about 500 light-years. By the early 1900s, astronomers were in a position to say that the stars lie at distances measured not just in dozens of light-years but hundreds upon hundreds of light-years. The Milky Way Galaxy, that vast collection of about 150 billion stars to which our Solar System belongs, was discovered to be a huge place. From edge to edge it measures about 100,000 light-years, and the Sun lies some 30,000 light-years from the center.

The Luminosity of Stars

The inverse square law of light also enabled astronomers to calculate the luminosity of stars. Here is an example of how finding the distance to a star by the parallax method and then applying the inverse square law of light tells us the star's luminosity.

The star Pollux has a parallax of 0.093 second and, therefore, must be 35 light-years away. With a photocell we can measure the apparent brightness of Pollux at its distance of 35 light-years. Say that the apparent brightness is 2 "brightness units." Now we can imagine Pollux to be only half its present distance from us, or 17.5 light-years. If it were, then its apparent brightness would be 8 brightness units, according to the inverse square law. And if it were only a quarter of its present distance from us, or 8.75

light-years away, it would be 16 times brighter than we now
see it to be, or 32 brightness units. We can keep imagining
Pollux to be closer and closer to us until it is right alongside
the Sun. So now we can compare its actual brightness with
that of the Sun since the two are side by side. When we do,
we find that Pollux is 36 times more luminous than the Sun.
So what is the luminosity of Pollux? Multiplying the Sun's
luminosity of 56×10^{26} cal/min by 36 gives us the answer:
$36 \times 56 \times 10^{26}$ cal/min $= 20 \times 10^{28}$ cal/min.

STAR	RELATIVE LUMINOSITY
Rigel	60,000
Deneb	58,000
Betelgeuse	4,000
Eta Aurigae	580
Zeta Leonis	50
Sun	**1.0**
61 Cygni B	0.028
Lalande 21185	0.0048
Ross 128	0.00030
Proxima Centauri	0.000052

It has been only in fairly recent times that astronomers
have found an answer to the age-old question: How bright
are the distant stars? Some stars are very much brighter
than the Sun, but most are very weak radiators of energy
compared with the Sun. For example, we can select nine
stars and compare their luminosities with that of the Sun.
While Pollux is 36 times more luminous than the Sun, notice

From a great distance in space our home galaxy, the Milky Way, would resemble this view of the Andromeda Galaxy some two million light-years distant. The two bright objects are smaller companion galaxies. The smaller dots are stars belonging to our galaxy. LICK OBSERVATORY

that Rigel is 60,000 times more luminous! Rigel is so luminous that if it took the place of the Sun, Earth would be vaporized almost instantly by the tremendous amount of radiation reaching it. At the other extreme, if Ross 128 were our local star, Earth would receive so little energy that it would be a deep-freeze planet.

So in our search for knowledge about the stars, we find that they show a tremendous range in luminosity.

The Size of Stars

The fact that we call certain stars red giants or blue giants and certain others red dwarfs or white dwarfs suggests that there is a large difference in size among the stars. But how do we know? No matter what its size, each distant star we observe appears only as a pinpoint of light through even the most powerful telescopes.

To find out the size of stars other than the Sun, astronomers closely examine the color of a given star. At first this may seem like a strange way to find out how large a star is. You know from your own experience that there is a direct relationship between color and temperature. If you place a cold poker in the hot coals of a fire, the poker heats up and eventually begins to glow a deep red. If you used bellows and heated up the coals still more, the poker would soon turn pink and then become white-hot. If you could make the coals even hotter, the poker would turn bluish-white as its temperature rose still higher.

Because there is this direct relationship between the color of an object and its surface temperature, astronomers can quite accurately assign surface temperatures to all the individual stars they can observe. The reddish stars we see—such as Betelgeuse or Ross 128—all have a surface temperature of around 3,000 K, no matter how large or small such a reddish star may be. Yellowish-white stars like the Sun have somewhat higher surface temperatures, up around 6,000 K. And the bluish-white stars, like Rigel, have surface temperatures of 50,000 K and more.

As you found in our comparison of the luminosity of 10 stars on page 83, Ross 128 is only 0.0003 as luminous as the Sun. You also saw that Betelgeuse is 4,000 times more luminous than the Sun. Both of these stars shine with the same reddish light, so both must have a surface temperature around 3,000 K. But how can we account for this extreme difference in luminosity between two stars that have almost exactly the same surface temperature? Because they do have the same surface temperature, each square-centimeter patch of Ross 128's surface must be emitting the same number of calories per minute as each square centimeter patch of Betelgeuse's surface.

The only way Betelgeuse can pour out vastly more energy than Ross 128 is for Betelgeuse to have more square centimeters of surface from which to radiate energy. In other words, Betelgeuse has to be a much larger star than Ross 128. That is the only way Betelgeuse can have a luminosity thousands of times greater than that of Ross 128. In a nutshell, that is how astronomers can figure out the size of a star by first examining the star's color.

Astronomers of ancient times could only guess that certain stars that seemed brighter than certain other stars probably were larger than the dimmer ones. But their guesses were more often wrong than right, since they had no idea of how far away a given star was. So here is another celestial puzzle that has been solved only in fairly recent times. Today we can look about us in the sky and point to red giant stars, blue giants, supergiants, and dwarf stars. Betelgeuse is one such giant star with a diameter some 400 times larger than the Sun's. If Betelgeuse replaced the Sun as our local star, it would swallow up Mercury, Venus, Earth, and fill up the Solar System out to the orbit of Mars. And Betelgeuse is not an especially large star. There is one star, a member of the double-star system called Epsilon Aurigae, that has a diameter 5,000 times

While the star Betelgeuse has a diameter 400 times greater than that of the Sun, the star Epsilon Aurigae has a diameter 5,000 times that of the Sun. (Star and planet size are not drawn to scale.)

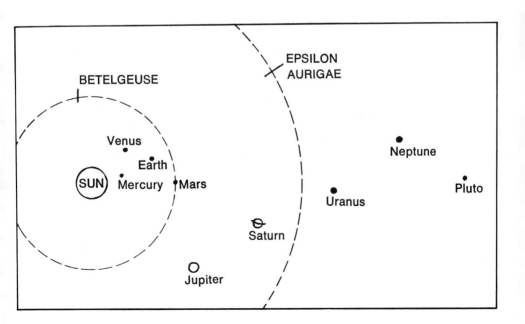

larger than the Sun's. That is a diameter measuring some 30 astronomical units! If Epsilon Aurigae were our local star it would fill up the Solar System to about halfway between Saturn and Uranus.

What about stars smaller than the Sun? How much smaller can they get and still be a star? Some of the white dwarf stars, which we'll return to later, are only about the size of Earth. Others are even smaller, measuring only about 1,000 km (620 mi) in diameter, which makes them smaller than the Moon. In a way, these smaller stars are far more interesting to astronomers than are the giant ones.

The Mass of Stars

Astronomers were able to calculate the Sun's mass once they knew the size of Earth's orbit and the length of time it takes Earth to circle the Sun once. By also studying the size and period of one star's orbit about its companion star in a double-star system, astronomers can calculate the mass of the two stars. And since about half of the stars we can observe are members of a double-star system, it is possible to compute the mass of a few stars by direct measurement.

But what about stars that are not double stars? How can astronomers say how massive those stars are? In the early 1900s, a British astronomer named Sir Arthur Eddington arranged numerous double stars in a diagram like the one shown here. He began by listing a large range of luminosities up and down one edge of the diagram. Then along the bottom he listed a wide range of stellar masses. The purpose of his *mass-luminosity diagram*, as it is called, was to

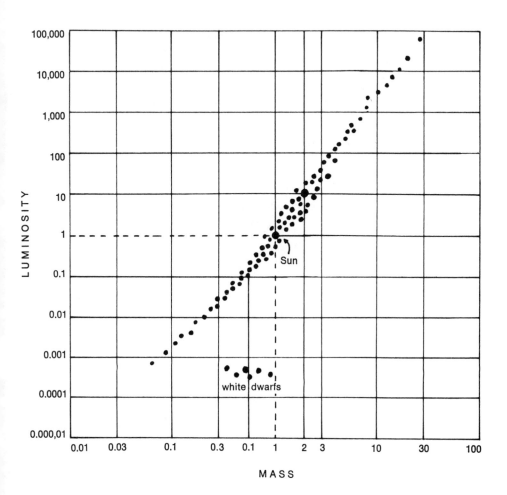

The mass-luminosity diagram enables astronomers to estimate the
mass of a star when they know its luminosity.

see if there might be some relationship between a star's luminosity and its mass. If he turned out to be right, he would then be able to say how massive a given single star was when its luminosity was known.

Notice that Eddington assigned the Sun a luminosity value of one and a mass value of one. Next, he began to position on his diagram numerous stars whose luminosities and masses were known. For example, a star 10 times more luminous than the Sun and 2 times as massive fell just above and to the right of the Sun's position. Another star 0.1 times as luminous as the Sun and only 0.3 times as massive fell below and to the left of the Sun's position.

When he had finished positioning many stars with known luminosities and masses on the diagram, he was delighted to find a neat pattern. Instead of being scattered all over the diagram, nearly all the stars fell along a diagonal line. There were only a few exceptions, all of which fell below the central part of the line. These were the mysterious white dwarf stars.

What did it all mean? Eddington reasoned this way: Stars whose luminosities and masses both were known fell along the diagonal line. Why, then, shouldn't stars only whose luminosities are known also fall along the same line? For example, an astronomer calculates the luminosity of a single star and finds that the star is 100 times more luminous than the Sun. He does not know the star's mass, because it is a single star. To find the star's mass, he locates 100 on the luminosity edge of the diagram and follows the 100-line straight across to the diagonal star-line and draws a dot there. He then follows the vertical line straight down to the mass edge of the diagram and finds that the star in question has a mass 4 times that of the Sun (or, $4 \times 2.0 \times 10^{33}$ grams $= 8 \times 10^{33}$ grams).

Eddington was able to make up two general rules about the relationship between stellar masses and luminosities. First, the more luminous a star is, the more mass it has. Second, the less luminous a star is, the less mass it has. Eddington had provided a way of estimating the masses of single stars whose luminosities were known.

There is a large range in the luminosity of stars. There also is a large range in the size of stars. But as the mass-luminosity diagram shows, there is only a small range in the mass of stars. Most stars that we can observe are only about 10 times more massive or about 10 times less massive than the Sun. But there are some exceptions with masses 50 times more or 50 times less than that of the Sun. These stars provide astronomers with many important clues about how stars are formed, what happens to them during their youth and middle age, and what kinds of fates await them.

We have come a long way since the time of the ancient Greek astronomers. With a knowledge of chemistry and physics, which they lacked, we can study the stars in a new light, in ways impossible for Hipparchus, Heraclitus, Empedocles, and the others. With our new knowledge, we no longer need to turn to the supernatural world to find answers to difficult questions. Questions that trouble us today—and there are many—may not be answered right away. Never mind, they will be answered tomorrow by an astronomer yet to be born.

9

THE BIRTH OF STARS

Where do stars come from? They have not always been shining as we see them now. Nor will they continue to shine forever as we see them now. As surely as they were formed, they are destined to end their lives as objects very different from the ones we see tonight.

The Stuff of Stars

If you scooped up 1 cc of air at sea level, and if you could count the number of molecules of gases making up that small parcel of air, you would count about 3×10^{19} particles, or 30,000,000,000,000,000,000 particles. But the higher we climb up through the atmosphere, the fewer such particles there are per cubic centimeter of space. In addition, but scattered around even more thinly, are simple compounds, about 30 of which have been identified so far. For example, they include methane (CH_4), ammonia (NH_3), water vapor (H_2O), and carbon monoxide (CO), among others. There is also matter called celestial "dust," which is of unknown composition but whose molecules are in the form of gas.

For the most part, space between the stars can be thought of as empty, since it contains so very little matter. However, in just about every direction in the sky where astronomers point their telescopes, space is either veiled or blocked by clouds of gas and dust called *nebulae*. One of the most splendid nebulae in the heavens is one called the Horsehead Nebula. It is classed as a *dark nebula* because it does not give off any light. We see it as a silhouette, its shape being outlined by the light of stars shining behind it. Another and more spectacular dark nebula is the dark band, or "rift," running part way along the summer Milky Way.

Certain other nebulae are classed as *reflection nebulae*, so named because their dust reflects the light of one or more stars imbedded in the nebulae. For example, the gaseous region around the Seven Sisters, or the Pleiades, resembles a cluster of streetlights on a foggy night. These stars are not especially hot stars, having temperatures less than about 25,000 K.

A third class of nebulae is represented by the Great Nebula in Orion, the most magnificent of these nebulae. The Great Nebula is said to be an *emission* nebula because its gas intercepts the light energy from nearby hot blue stars, in this case four, with temperatures greater than about 25,000 K, and re-emits the energy. As a result, the gas cloud glows with its own energy. Like the other nebulae, the Great Nebula appears to be a dense cloud of gas and dust. But these nebulae actually are very thin. Their densities are far less than even the best vacuum we can produce in the laboratory. In the case of the Great Nebula, its density probably is not more than about 1,000 atoms and dust particles per cubic centimeter.

Planetary nebulae are still another class of these objects. One of the most famous is the Ring Nebula in the constellation Lyra. It looks like a giant smoke ring floating in space. Actually it is a huge shell of gas enclosing an intensely hot star. Stars that cast off matter and produce planetary nebulae are the hottest stars we know of and have surface temperatures on the order of 100,000 K. The gas of all emission nebulae glows with a fluorescent light at a temperature of about 10,000 K. Because this gas re-emits energy, the planetary nebulae also are emission nebulae. They were called ''planetary'' nebulae because they are spherical and greenish, and when they were first discovered in the 1700s they were mistaken for planets within the Solar System. But later, when it was shown that they do not ''wander'' among the stars as the planets do, astronomers realized that they were not planets but mysterious objects out among the stars themselves. There may be about 50,000 planetary nebulae in the Galaxy, and they may form at the rate of two or so a year. Depending on the class of nebulae we are considering, their life-spans may last from tens of thousands of years to billions of years.

Another class of nebulae is that caused by extremely high-energy stars that end their lives in catastrophic explosions—the supernovae, which we will examine in detail later. Two spectacular examples are the Crab Nebula in the constellation Taurus and the Veil Nebula.

Although nebulae in general are not very dense objects,

they do occupy large volumes of space and do contain a very large amount of matter. For example, let's imagine a "typical" nebula cloud containing about 1,000 particles of gas and dust per cubic centimeter and occupying a volume of two cubic light-years of space. In that volume of space, which isn't very large when you consider that the diameter of the Galaxy is about 100,000 light-years, there would be 2×10^{57} gas and dust particles.

What we want to find out is the total mass of matter in our typical nebula. So far, we know the number of gas and dust particles. To find out how many grams of star-stuff occupy our two cubic light-year volume of space, all we do is divide the total number of particles (2×10^{57}) by the number of particles it takes to make up one gram of matter, which is 6×10^{23} particles. The answer turns out to be 3×10^{33} grams of star-stuff. And that just happens to be close to the Sun's mass. So a typical nebula occupying a volume of space measuring two cubic light-years contains enough matter to form a star as massive as the Sun.

Let's return to the Great Nebula in Orion, which you can see with the naked eye and which shows up quite well through binoculars. That nebula seems to contain an average of about 1,000 particles of gas and dust per cubic centimeter. Since it occupies a more or less spherical volume of space about 20 light-years in diameter, the nebula contains about 7×10^{36} grams of gas and dust particles. That is about 3,000 times more mass than the Sun has. So if star formation began in the Great Nebula, and if all of the gas and dust forming the nebula were used up, more than 3,000 Sun-like stars could be formed.

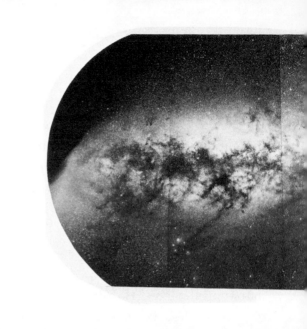

This is a composite photograph of the Milky Way as seen in the summer sky. The dark band, or "rift," is a dark nebula composed of dense concentrations of gas and dust. HALE OBSERVATORIES

The Horsehead Nebula, one of the most splendid visible to us, is an example of a dark nebula. We see it outlined by light emitted from stars on the far side of the nebula. HALE OBSERVATORIES

The Pleiades are an example of a reflection nebula, one whose
dust reflects the light of one or more stars embedded in the
nebula. LICK OBSERVATORY

The Great Nebula in the constellation Orion is an example of an emission nebula, one whose gas reemits the energy of nearby hot blue stars. LICK OBSERVATORY

This nebula in the region near the star 12 *Monocerotis* is another example of an emission nebula. YERKES OBSERVATORY

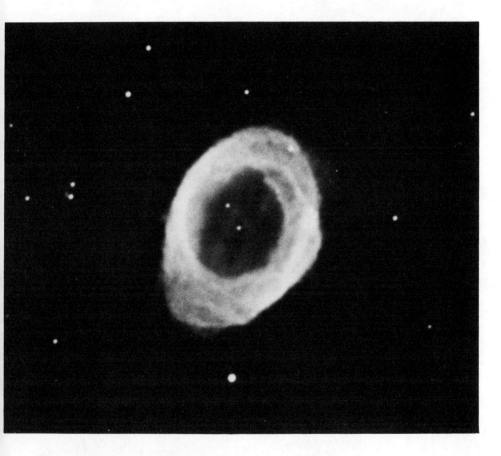

The so-called planetary nebulae also are emission nebulae. The
Ring Nebula in the constellation Lyra shown here is typical of
these roughly sphere-shaped objects. HALE OBSERVATORIES

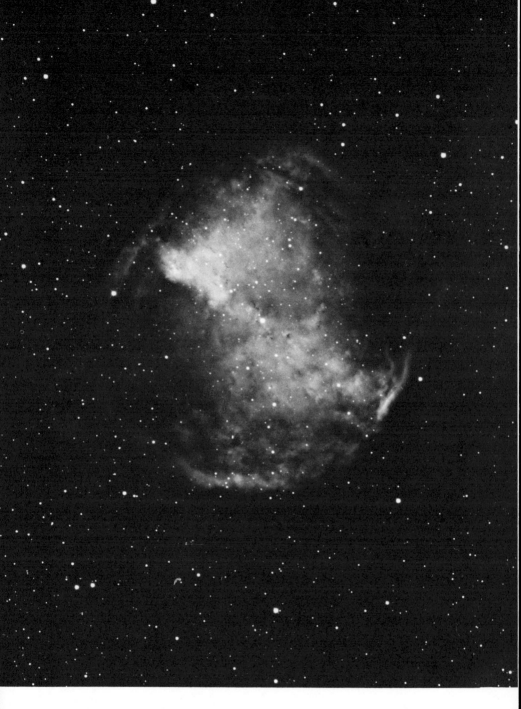

The Dumbbell Nebula is another example of a planetary nebula.

LICK OBSERVATORY

The Crab Nebula in the constellation Taurus is an expanding
cloud of gas, the supernova remains of a star observed by Chinese
astronomers in the year 1054. HALE OBSERVATORIES

The Veil Nebula in the constellation Cygnus is the remains of a supernova explosion that took place thousands of years ago. Such supernova remains cannot retain their identity for much more than a hundred thousand years. HALE OBSERVATORIES

How Stars Are Formed

The nebulae seem to be the source matter for star formation. At least astronomers cannot think of a better place to look. If the nebulae are the birth-grounds of stars, then shouldn't we be able to see some evidence of star formation in certain nebulae, assuming that stars are still being formed in the Galaxy?

Astronomers have certain clues that stars are now being formed in many places throughout the Galaxy and that those places are the nebulae. The clues are particularly dense and dark concentrations of gas and dust called *globules*. Newton's law of gravitation tells us that such dense concentrations of matter in space will attract more and more surrounding matter into themselves and so grow in size and density. We can imagine a continual infall of material toward the denser central region of a globule. As a result, the number density of particles in the interior of a globule must continually increase. And with the increase in number density, there must be an accompanying increase in pressure and temperature.

So depending on the size of the right kind of nebula for globule formation, one or more globules form and grow. They are the distant beginnings of a star, called a *protostar*. As compaction of material in the core region of a protostar continues, the new star radiates feebly, eventually glowing a deep red as its temperature rises. At this stage in the star's new life the nuclear furnace has not yet been ignited in the core, and it will not be until after the core temperature rises to about 10,000,000 K. When that time

Numerous globules, which are small, dark concentrations of matter thought to be the first stage of star formation, can be seen in a section of the Rosette Nebula. HALE OBSERVATORIES

Several globules can also be found in this photograph of a dark
nebula in the constellation Serpens. LICK OBSERVATORY

arrives, the protostar has achieved "adult" status, since hydrogen nuclei in the core are colliding vigorously and frequently enough to sustain an outflow of energy by the fusion of hydrogen nuclei into helium.

When that happens, a new star has evolved from a state of chaos, in which gas and dust particles at first were moving about in a state of disorder, into a highly ordered state of matter. And the star will carry on in a more or less steady-state condition for millions to many billions of years. How stable it remains, and for how long, depends on one important condition—the amount of mass it acquired when it was forming as a globule.

The Sun is slightly more massive than most of the stars we can observe. The most massive stars we know of are the *blue giants* like Rigel, Bellatrix, and the Orion Nebula stars. Rigel is an easy star to pick out. It appears very bright to the eye and is found in the lower right-hand corner of the constellation Orion.

The blue giants are tens of times more massive than the Sun. Rigel, for example, is 20 times more massive and some 60,000 times more luminous. The cores of these stars have temperatures of more than 20,000,000 K and surface temperatures of 50,000 K and more. It is because of their high surface temperatures that we see these stars shining with a bluish-white light.

Because Rigel and the other blue giant stars are so massive, you might think that they have the longest life-spans, much longer than the Sun's life-span of about 10 billion years, for example. It turns out that the blue stars have

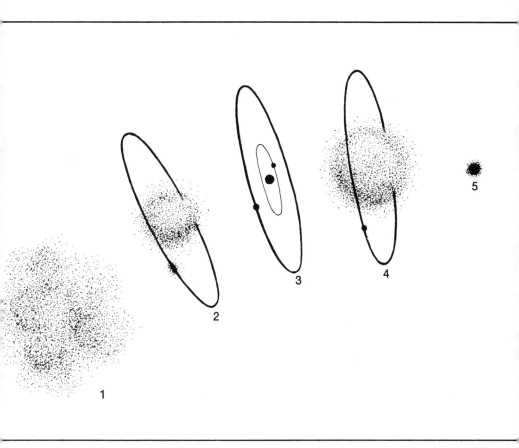

Stars are thought to form out of clouds of gas and dust. First a globule forms and condenses into a dense sphere of gas (1), which heats up and eventually emits energy by fusing hydrogen nuclei in its core region (2). When the star eventually exhausts its fuel supply (3), it swells up as a red giant (4). In its dying stages it contracts and becomes a white dwarf (5).

exceptionally short life-spans. They are spendthrift stars. Although such a star may have 20 times more mass than the Sun, it is using up its hydrogen fuel at an enormous rate. Rigel, for example, is pouring out so much energy, and as a result is burning up its hydrogen fuel supply at such an enormous rate, that the star has a life span measured in millions rather than billions of years!

Perhaps now you can guess the future of those globules that evolve into stars of low mass and become *red dwarf stars*. Ross 128, for example, is only one-tenth or so as massive as the Sun. Although it acquired enough mass to ignite the nuclear furnace in its core, it did not acquire enough to push the core temperature much higher than the ignition point of about 10,000,000 K, and the surface temperature to not quite 3,000 K. So Ross 128 is but a feeble radiator of energy and is only 0.0003 times as luminous as the Sun. But because red dwarf stars like Ross 128 are such conservative users of their hydrogen fuel, in spite of their small mass they have life-spans measured in the trillions of years.

10.

THE DEATH OF STARS

How Stars Go Out

Because stars evolve over such very long periods of time, we have no hope, at the present at least, of ever recording what happens throughout a star's life-span. But perhaps such records do exist, made in patchwork fashion through time by various advanced technological civilizations scattered about the Galaxy.

Although we cannot observe what happens to a given star over its entire life-span, we can do the next best thing. We can observe many stars at various stages of their life-spans and so gain at least some idea of what may happen to a star from beginning to end.

One thing we know for certain is that no stars can go on shining forever. One day the star must stop shining as a steady-state star and enter its final stage when it has used up its core hydrogen. During the time when hydrogen nuclei were fusing and building up helium nuclei, vast amounts of

energy were flowing outward from the intensely hot core
region of the star. It is this outflow of energy that keeps a
star puffed up and prevents it from collapsing in on itself.

Now try to picture what happens to a star whose hydrogen supply is running low. As the hydrogen nuclei in the
core become fewer and fewer, there are fewer collisions and
so fewer fusions for energy production. As the outward flow
of energy from the core weakens, it eventually reaches the
stage when it is so feeble that the inner part of the star collapses in on itself. When this happens, the temperature and
pressure in the core region suddenly increase to such high
values that hydrogen nuclei just outside the core begin
fusing. Higher temperatures in the core also mean that all
those helium nuclei that had been built up from hydrogen
fusions over billions of years themselves can fuse and form
the nuclei of still heavier atoms, in this case carbon. The
high core temperature next causes the carbon nuclei to fuse.
In very massive stars, iron eventually is produced. Elements heavier than iron cannot be manufactured in aver-
age-type stars such as the Sun, only in supernova explosions.

This new outburst of energy resulting from the fusions of
nuclei heavier than hydrogen causes the outer layers of the
star to swell up as its gases at and near the surface are
pushed outward by the increased pressure in the core.
Because the star now has a much larger surface area than
before, it no longer shines with a yellowish-white light. The
energy pouring up from the core region is now spread out
over a much larger surface area of the swollen star. This
means that we now see the star shining with a cool, red

light. The star has evolved into a *red giant* and will remain that way for several millions of years.

Eventually, as the helium and carbon nuclei in the star's core are used up, fusions usually cease, although some stars fuse oxygen, silicon, and other materials. Again, there is a decrease in energy outflow from the core region and the core again collapses in on itself. But this time no new fusions can occur. The entire star now takes part in the collapse and tumbles in on itself until it is an object only about 100-thousandths of its size when a steady-state star. About 5 billion years from now the Sun is due to swell up as a red giant, at which time it will become so large that it will engulf Mercury and Venus. Then when it collapses, it will shrink to a relatively tiny object about the size of Earth, or smaller.

At this shrunken and highly condensed stage in a star's life, it is still giving off vast amounts of energy, but not through nuclear fusion. Its energy output now comes from the tight compaction of material around the slowly cooling core. The star is now so dense that a teaspoon of it would weigh several tons. Because there is now such a small surface area from which the star's dwindling store of energy can be radiated, the star is seen as an intensely white object. It has become a *white dwarf star*. A white dwarf cannot go on radiating energy forever, since no new energy is being produced in the core. Its nuclear furnace has been shut down for good. Gradually the star cools, giving off less and less energy over a period of millions of years more. Eventually its light fades and it becomes an object that can best be described as a *black dwarf*.

Since black dwarfs would not be radiating energy, they

would be difficult objects to detect. In fact, astronomers have invented black dwarfs; they are only a theory. But it would seem that such relics of stars must exist out there in the dark. Astronomers cannot come up with a better explanation of what happens to white dwarfs, which are known to exist.

Supernova Stars

Earlier you found that the stars are the element factories of the Universe. And you saw how a star about as massive as the Sun is able to produce the nuclei of atoms up to the weight of helium, and possibly carbon. Only very massive stars manage to reach the iron stage, those rare monstrosities of stars that become *supernovae.*

Stars like the Sun, along with the red dwarfs, do not have enough mass to become supernovae. It is the blue giants that evolve into these explosive stars that hurl off massive amounts of matter and so enrich interstellar space. The first supernova we have any record of is the Guest Star observed by Chinese astronomers in the year 1054 and shown on page 103. Today we see its remains as the Crab Nebula, a cloud of cast-off material about 42 light-years in diameter and expanding at a rate of 1,300 kilometers (800 miles) a second. Speeds that great—compared with a nebula's average speed of about 24 kilometers (15 miles) per second through space—can be triggered only by a gigantic explosion.

At the center of the Crab Nebula we can observe a very odd star, but it was not until the late 1960s that astronomers discovered just how odd the star is. Earlier, in the 1940s, astronomers had found that this star's surface temperature far exceeded that of even the hottest blue giants. They estimated a surface temperature not of 50,000 K but of hundreds of thousands of degrees! But more about this mysterious star in the center of the Crab Nebula in a moment.

Here is how astronomers now think a supernova explosion is triggered. Picture a blue giant star that has gone through the red giant stage and has collapsed in on itself. The heavy nuclei that had been built up in the star's core can no longer withstand the continually increasing crushing force. Eventually they are crushed so forcefully that they begin to break apart into individual protons and neutrons, the two basic building blocks that make up the nuclei of all atoms heavier than hydrogen. Because a proton has a positive electrical charge, two protons push each other apart, as do two like poles of a pair of magnets. Because neutrons are electrically neutral, these particles do not repel each other, and they are not repelled by protons. To make a complete atom, all you have to do is add one or more electrons to a nuclear package of protons and neutrons. The electrons are held captive of the atom because they have a negative electrical charge and so are attracted to the protons. But in the core of a star any whole atom would be knocked about with such force that it would lose its electrons. The electrons would swim about independently, leaving only the nucleus of the atom intact.

But again, the crushing forces in the core of a supernova are so great that the nuclei break apart. There is a "soup"

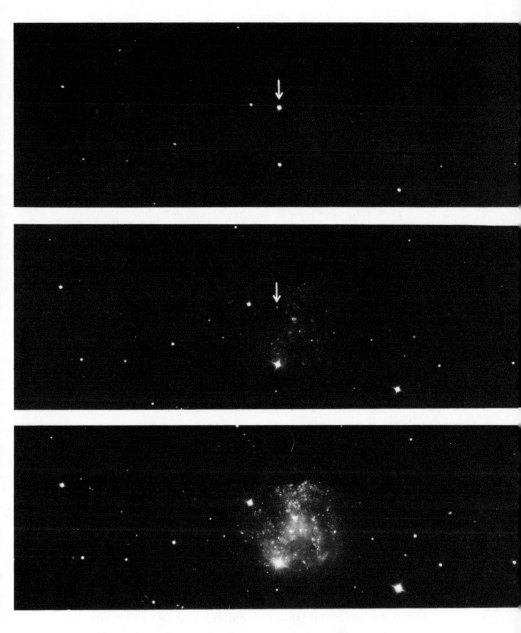

Three views of a star that became a supernova. The top view shows the star at maximum brightness in a time exposure of 20 minutes, August 23, 1937. The middle view shows the same star on November 24, 1938, when it had become very faint, in a time exposure lasting 45 minutes. In the bottom view, taken January 19, 1942, the star is too faint to be observed even in a time exposure lasting 85 minutes. The longer exposure is responsible for the appearance of additional stars. HALE OBSERVATORIES

of free-swimming electrons, protons, and neutrons. The enormous number of these free subatomic particles produces a catastrophic explosion. The star literally blows off its upper layers, leaving the intensely luminous core region exposed. Try to imagine such an explosion, a blinding burst of light brighter than 100 billion Suns, or about as bright as an average galaxy. Some supernova explosions observed in other galaxies are brighter than the entire galaxy containing the exploding star. Over the past 90 years, more than 400 supernovae have been observed in galaxies other than our own.

Astronomers recognize two types of supernovae. Those of Type I are somewhat more massive than the Sun and seem to be triggered when carbon in the supernova's core begins wholesale fusion into heavier elements. Type I supernovae flare up to maximum brightness over a period lasting several days. They maintain their brightness for several days more, then over a period of about a month decline in brightness.

Type II supernovae involve the most massive stars, those with cores of iron nuclei. They are found mostly in the arms of spiral galaxies where there is lots of gas and dust and many massive, young stars. When the core temperature of a Type II supernova reaches a certain value, the iron nuclei become unstable and the core collapses. Outer regions of the star also collapse and send the core temperature soaring still higher. All remaining nuclear fuel in the core fuses explosively. Type II supernovae rapidly build in brightness as the Type I stars do, remain bright for several days, and then decline somewhat in brightness over about 25 days.

Then, unlike the Type I supernovae, the Type II stars do not fade from view. Instead they remain rather bright for between 50 and 100 days. After that, a Type II supernova's decline is rapid.

Neutron Stars

When we look at the Crab Nebula, we are looking at the exposed core of what was a blue giant star now surrounded by a vast cloud of ejected upper-layer material. Eventually, what is left of a supernova is a huge ball of neutrons. Since neutrons are electrically neutral, they can be packed very closely together, because they do not repel one another. This means that such a star is fantastically dense. Called a *neutron star*, it is compressed into a ball only about 16 kilometers (10 miles) in diameter, although the star contains as much matter as the Sun with its diameter of 150 million kilometers (850,000 miles). A lump of neutron star matter the size of a suger cube weighs 10 million tons, making a similarly sized lump of white dwarf matter seem like a marshmallow by comparison.

Pulsars

For several years neutron stars were only a theory. Like black dwarfs, they were an invention of astronomers. But then in 1967, British astronomers working with a radio telescope picked up some unusual signals from a star. The signals were pulses of energy so regularly spaced that one of the astronomers half jokingly named the object emitting the signals "LGM," standing for "Little Green Men." In other words, some of the astronomers on the listening project did not rule out the possibility that the signals were being beamed toward us by intelligent beings of an advanced technological civilization.

It turned out that the signals were not being broadcast by little green men. Instead they were being transmitted by a rapidly rotating neutron star, a class of stars called *pulsars*. When the core of a massive blue giant star collapses in on itself at the end of the red giant stage, whatever rotation the core of the star had initially is multiplied many times. The greater the collapse, the faster the star rotates.

The reason a rotating neutron star emits signals is that the star's magnetic field rotates right along with the star itself. So there is a gigantic rotating magnet in the sky, a magnet that generates electromagnetic waves that our radio telescopes can detect. The remains of the Guest Star seen in the middle of the Crab Nebula are a pulsar, and the most rapidly rotating one discovered so far. It sends out about 30 pulses a second, but it is slowing down at the rate of 38-billionths of a second each day.

A slowly rotating star like the Sun, which rotates only once every 25 days, takes on the shape of a sphere. But the rapidly rotating pulsars bulge out at the equator and take on the shape of a doorknob. This happens because centrifugal force tends to cast off matter around the star's equatorial region. As a pulsar's rate of rotation slows, we would expect the star gradually to lose its doorknob shape and become more spherical. But can it, since the outer mile or so of a neutron star probably is a rigid shell billions of times more rigid than steel? It apparently can, and each time it does there must be a violent starquake.

This is not the end of a blue giant's life history though. As a pulsar, its rotation slows until it is rotating so slowly that it again undergoes gravitational collapse. But this collapse does not span many years. Instead it may take place suddenly, in less than one-hundredth of a second. This super-dense collapsed star with a super-rigid shell is now so dense that a chunk the size of a sugar cube would weigh a billion tons.

Black Holes

The pull of gravity at the surface of one of these super-dense stars is so strong that no energy whatever can escape from the star—not even light. We will, therefore, never be able to see one of these fascinating objects, since it would simply be a black "hole" in space. And that is what astronomers call those massive blue giant stars that evolve into red giants, then explode as supernovae, and finally collapse into oblivion as *black holes*. There is some argument, however, whether neutron stars become black holes. They may be two quite different products of stellar evolution, depending on the original mass of the star.

Astronomers "invented" black holes, just as they invented black dwarfs and neutron stars. Although some astronomers deny that black holes exist, others insist that black holes have been detected, even though we can't see them. Since the gravitational attraction of a black hole is so strong, anything in the region of one of these incredible objects will be pulled into it. How many black holes are there? No one can say. It now seems that Cygnus X1 is a black hole, and there are several other candidates.

Nova Stars

Although the supernovae are the most spectacular stars that cast off matter into space, there are still other kinds. Stars called *novae* are lesser versions of the supernovae. They flare up suddenly and mysteriously, increasing in brightness over a period of days or several weeks. They then return to normal again. An example of a nova is the one called Nova Hercules, shown here. Maybe 50 or so novae flare up in the Galaxy every year.

Nova stars are stars that suddenly increase in brightness and then return to normal. Shown here is Nova Hercules, as seen on March 10, 1935, after it brightened as a nova (left), then again two months later on May 6 (right) after it had returned to normal. LICK OBSERVATORY

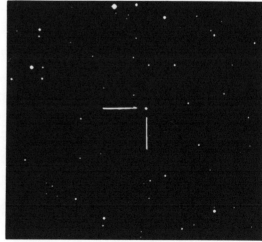

Shell Stars

Shell stars are still another class of stars that explosively eject matter off into space. The one shown here is in the constellation Aquarius. It appears to be a ring around the central star. Actually it is a shell only appearing as a ring because we are looking through a greater thickness of matter out around the edge of the shell than when we look through the shell to its center. This particular shell of gas is expanding at the rate of about 16 kilometers (10 miles) each second. What makes shell stars explosively eject matter off into space is still a mystery.

Variable Stars

There is a large class of stars called *variable stars* because their energy output varies. Most of these stars are in advanced stages of their lives. Of the more than 100 billion stars in the Galaxy, variable stars seem to be uncommon. About 18,000 or so have been discovered to date. Shell stars, novae, and supernovae all are explosive variables. In contrast, there also are pulsating variables that go through regular cycles of brightening and dimming.

Most of the known pulsating variable stars are of the Mira type, named after the first such star discovered, Mira, in the constellation Cetus. Mira-type variables are almost always invisible except through a telescope. Their typical *period*, or time of completing one cycle of going from dim to bright and back to dim again, is 300 days. During a period, a Mira-type variable star becomes about 15 times more

This planetary nebula, known as NGC 7293, in the constellation
Aquarius is an example of a shell star. HALE OBSERVATORIES

luminous than when it is at the least luminous part of its cycle. All variable stars with long periods of the Mira-type are red giant stars with surface temperatures ranging from about 1,900 K to about 2,600 K. Nearly one-quarter of the total number of known variables in our galaxy are Mira-type.

Another large class of pulsating variable stars is the RR Lyrae variables, named after the first such variable star discovered, in the constellation Lyra. We know of a little more than 3,000 of them associated with our home galaxy. The pulselike brightness beat of the RR Lyrae variables is much shorter than the period of Mira-type variables, taking from 6 to 18 hours to complete one period. One thing that is interesting about the RR Lyrae variables is that all of them seem to have the same luminosity. Suppose an astronomer spots an RR Lyrae variable in another galaxy. He knows it is an RR Lyrae star by measuring its period. Knowing its luminosity, the astronomer can next measure its apparent brightness. Then by applying the inverse square law of light, he can say how far away the galaxy is. The RR Lyrae variables scattered about the Galaxy, and contained in globular clusters of stars forming a halo around the Galaxy, enabled astronomers earlier in this century to determine the size of the Galaxy. It would be difficult to overstate the important role the RR Lyrae variables have played in our working out distances to various parts of the Universe. The RR Lyrae variables are white or yellow-white giant stars.

The best-known, but not the most numerous, pulsating variables are the Cepheid variables, named after the first

This globular cluster of stars seen in the constellation Hercules is
typical of other such clusters forming a halo around the nucleus
of the Galaxy. HALE OBSERVATORIES

such star observed, in the constellation Cephius. We know of nearly 1,000 of them. The Cepheid variables are among the largest stars in our galaxy and have periods ranging from a few days to about 50 days. These are extremely luminous stars. Like the RR Lyrae variables, the Cepheid variables can be used as metersticks to measure distances within the Universe, because there is a precise relationship between their period and average luminosity. In 1924, the American astronomer Edwin Hubble used Cepheid variables in his work that led to the discovery that the Andromeda ''nebula'' actually was a galaxy lying some 2 million light-years beyond our own star system.

None of the variable stars—either the eruptive (explosive) or pulsating kind—is well understood. Astronomers still cannot say what makes a pulsating star pulsate or a nova go on a rampage. They now think that the Cepheids and other pulsating stars begin their adult lives in quite a normal way, by generating energy through the fusion of hydrogen nuclei when the core temperature of the protostar reaches about 10 million K. Then at one stage in life they develop some kind of ''nuclear ailment'' and begin to throb. After perhaps a few million years the star ''recovers,'' stops pulsating, and once again shines at a steady-state rate.

From Dust to Dust

Shell stars, novae, supernovae, and pulsating variables all cast off matter into space. But the matter ejected is different from the original hydrogen-matter out of which the star was formed. Particularly in the case of supernovae, the ejected secondhand matter contains many elements heavier than hydrogen. In dense enough concentrations, this second-hand matter can form globules and evolve into new stars. Such stars are said to be *Population I* stars, as opposed to *Population II* stars, which are formed out of hydrogen. The Sun is one such Population I star formed out of secondhand matter more than 5 billion years ago.

Is there some special significance to Population I stars? Before answering that important question, let's consider what may happen to our galaxy a few billion years from now. Stars are born out of the gas and cosmic dust of space, emit energy for a few million or billion years through the process of nuclear fusion, and then evolve into black dwarfs or black holes, burnt-out objects that can no longer be seen. If all we have said so far in this book about stellar evolution is true, then there must be many "dead," invisible stars out there secretly wheeling about the galactic hub. As we can imagine a time several billion years ago when our planetary system, or any other planetary system for that matter, was a disk of gas and dust, we can also imagine a time when the Galaxy itself had not yet given birth to stars, but itself was a vast disk of gas and dust.

From what astronomers now know about star formation, it seems likely that the oldest stars in the Galaxy are those Population II stars making up the globular clusters and the RR Lyrae variables. Those stars are now thought to be about 10 billion years old, or twice the age of the Solar System. Although some stars certainly are now being formed in the spiral arms of the Galaxy, the great age of star formation is far behind us. That must be so because practically all of the matter in the Galaxy is now locked up in the stars themselves. Only one percent of the total matter contained in the Galaxy now seems to exist as interstellar gas and dust. As more and more of this one percent of matter forms globules that evolve into stars, fewer and fewer stars can be formed.

It would seem that eventually a time will come when star formation must cease. When that day arrives, we will be members of a dying galaxy. Astronomers viewing us from distant galaxies will see ours as an aging galaxy shining with a reddish glow generated by numerous red giants and red dwarf stars. Eventually the lights will go out, one by one, as each dying star emits energy so weakly that it can no longer be detected across space. When that time comes, our galaxy will be a dark galaxy, a vast and silent grave-yard of some 150 billion phantom objects invisibly wheeling about the Galaxy's center. The future of the Galaxy is necessarily tied to the evolution of its individual stars. Are there phantom galaxies lying hidden out there in the dark now? We do not know. It may be that the Universe is not yet old enough to have permitted dark galaxies to evolve.

What about the question posed earlier—"Is there some special significance to Population I stars?" Indeed there is. Earth, Mars, and the other planets of the Solar System are composed of the same material out of which the Sun originally was formed: lots of hydrogen, lesser amounts of helium, and still smaller amounts of heavier elements such as carbon, oxygen, iron, and the many others contained in the Solar System nebula of Population I matter, the debris cast off by worn-out stars.

Since all living organisms on Earth originated out of those atoms and molecules of which Earth itself was originally formed, then your body must contain atoms cast off from stars more than 5 billion years ago. Some time in the distant past the same heavy atoms now in your body took part in a reaction deep within the core of a supernova star. A small part of you once helped that star shine for a fraction of a second of its lifetime. You and I are products of stellar evolution. For a brief time our atoms will remain organized in a living biological system. Eventually, though, long after Earth life has been snuffed out by a red-giant Sun, our atoms may once again be released to space, and once again become involved in stellar evolution, and once again give rise to life in a never-ending cycle of birth, death, and rebirth.

GLOSSARY

Absolute Temperature Scale   A temperature scale based on the concept that all atomic motion ceases at a temperature defined as absolute zero. One such scale is called the Kelvin temperature scale, after its developer, Lord Kelvin. To convert from the Kelvin scale to the Celsius scale, subtract 273 from the Kelvin reading. (See also Celsius and Fahrenheit temperature scales.)

Absolute Zero   The temperature at which all molecular motion was presumed by Lord Kelvin to stop. On the Kelvin temperature scale, absolute zero is written as 0 K and is equal to $-273°C$.

Apparent Brightness   The measure of a star's observed brightness; how bright a star appears to the eye as opposed to its actual brightness, or luminosity. The farther away a light source is from the observer, the less its apparent brightness will be, although its luminosity does not change.

APPARENT MOTION   The motion of any celestial object as seen from Earth, which itself is moving.

ASTEROID   Any of millions of rock-metal fragments ranging in size from a fraction of a meter to several meters across and traveling about the Sun in orbits lying between Mars and Jupiter.

ASTROLOGY   A superstitious belief that the relative positions and motions of stars and planets bring good fortune or ill fortune to people, animals, plants, nations, and institutions. The science of astronomy grew out of astrology more than 2,000 years ago.

ASTRONOMICAL UNIT (A.U.)   A measure of distance equal to the average distance of Earth from the Sun, which is established by the United States Naval Observatory as 149,600,000 kilometers (92,752,000 miles).

ASTRONOMY   The science dealing with celestial bodies, their distances, luminosities, sizes, motions, relative positions, composition, and structure. The word comes from the Greek and means the "arrangement of the stars."

BINARY STARS   Two stars held in gravitational association with each other and revolving around a common center of mass. Also called "double stars." Some star systems, such as the one to which Alpha Centauri belongs, have three or more stars held in gravitational association and are known as multiple-star systems.

BLACK DWARF   A star that has passed through the white dwarf stage and is radiating energy so weakly that it can no longer be observed directly.

BLACK HOLE   The incredibly dense remains of a star. Black holes are thought to be so dense that no radiation leaves them.

BLUE GIANT   An especially massive, large, and luminous star, such as Sirius V and Rigel I, which is seen to shine with a bluish-white light. The core temperatures and surface temperatures of these short-lived stars are many times higher than those for less massive stars such as the Sun.

CALORIE   The amount of energy required to raise the temperature of one gram of water one degree Celsius.

CELSIUS TEMPERATURE SCALE   The temperature scale on which ice melts at 0° and water boils at 100° at standard atmospheric pressure. Developed by the Swedish astronomer Anders Celsius in the eighteenth century. One Celsius degree is equivalent to nine-fifths of a Fahrenheit degree. To convert from Celsius to Fahrenheit degrees, multiply the Celsius value by 9, divide by 5, and add 32. Also called the centigrade temperature scale.

CEPHEID VARIABLE   A pulsating variable star with a brightness period of from one day to about 50 days. Because there is a relationship between the period and luminosity of Cepheid variables, these stars can be used as celestial metersticks to determine the distance to other galaxies.

CONSTELLATION   The grouping into imaginary figures of certain stars on the celestial sphere. The ancients recognized the groups as human and animal figures; for example, Orion "the Hunter" and so on. By international agreement, astronomers recognize 88 constellations.

DEGREE   One of 360 equal units composing a circle. Degrees
are subdivided into minutes, a degree containing 60 minutes. Minutes are further subdivided into seconds, a minute
containing 60 seconds. Also a unit to measure temperature.
(See also Kelvin, Celsius, and Fahrenheit temperature
scales.)

DENSITY   Mass per unit volume, or the amount of matter
contained in a given volume of space, and expressed as
grams per cubic centimeter. Water, for example, has a density of 1 g/cc.

DIAMETER   A straight line passing through the center of a
circle or sphere and extending from one edge to the opposite edge.

DOUBLE STAR   See Binary Stars.

ECLIPSE   The partial or total blocking from view of one
celestial object by another passing in front of it. A lunar
eclipse occurs when the Moon passes through Earth's
shadow. A *partial solar eclipse* occurs when the Moon blocks
only part of the Sun from view. In a *total solar eclipse*, the
Moon completely covers the Sun's disk. An *annular solar
eclipse* occurs when the Moon covers all the Sun's disk but,
because the moon is at its greatest distance from Earth, it
does not appear quite so large as the Sun and so leaves a
narrow rim of the Sun visible.

ECLIPTIC   The path the Sun and Earth appear to travel
across the sky in one year. It forms a great circle on the
celestial sphere.

ELECTRON   A negative unit of electricity and part of all atoms. Clouds of electrons surround the nuclei of atoms. The mass of an electron is 1/1,840 that of the hydrogen atom.

ELEMENT   A substance made up entirely of the same kind of atoms. Such a substance cannot be broken down into a simpler substance by chemical means. Examples are gold, oxygen, lead, chlorine.

ENERGY   That property of an object enabling it to do work.

EXPLOSIVE VARIABLE   A star such as a nova or shell star that suddenly varies in brightness. (See also Nova and Shell Star.)

FAHRENHEIT TEMPERATURE SCALE   The scale on which ice melts at 32° and water boils at 212°. Nine Fahrenheit degrees are equivalent to 5 Celsius degrees. To convert from Fahrenheit to Celsius, subtract 32 from the Fahrenheit reading, multiply by 5, then divide by 9.

GALAXY   A vast assemblage of stars, gas, and dust held together gravitationally. *Spiral galaxies,* the brightest of all galaxies, have a dense nucleus with less dense spiral arms. Our galaxy is a spiral galaxy. Because the galaxies are oriented every which way in space, we see some edge-on and others from all different angles. *Barred spiral galaxies* have arms that wind outward from the ends of a central bar. Some of them have arms that sweep around so that they nearly touch both ends of the central bar. Barred spirals seem to have unusually large amounts of gas and dust. *Elliptical galaxies* are slightly flattened sphere-shaped galaxies. While some are very elongated, others form nearly perfect globes. We cannot see any structure in these star systems, nor is there evidence that they contain large clouds of gas and dust. *Irregular galaxies* are so named because they have no regular shape. The nearby galaxies, the Clouds of Magellan, are both irregular galaxies.

GLOBULAR CLUSTER   A globular assemblage of some 100,000 or more stars. A halo of about 100 globular clusters forms a sphere around the central part of our galaxy.

GLOBULES   Especially dense concentrations of gas and dust that appear to be the first stages of star formation. Globules have been identified in several large nebulae.

GRAVITATION   The force of attraction between any two or more objects in the Universe, no matter how large or small. The attraction between any two objects in the Universe is proportional to their mass and inversely proportional to the square of the distance between them. The greater the mass, the greater the force of attraction; the greater the distance, the less the force of attraction.

HALF-LIFE   The period of time during which one half the number of atoms of a radioactive element change into atoms of a different element. (See also Radioactivity.)

KELVIN TEMPERATURE SCALE   See Absolute Temperature Scale.

KINETIC ENERGY   Energy of motion; for example, of the atomic particles making up a gas, as they collide with each other and against the walls of their container.

LIGHT-YEAR (L.Y.)   The distance that light travels in one year, at the rate of 299,000 kilometers (186,000 miles) per second, which comes to about 10 trillion kilometers (6 trillion miles).

LUMINOSITY   The total amount of radiation emitted by an object. Astronomers generally express a star's luminosity in units called "ergs," but in this book calories are used because of the activity involving the measurement of the Sun's energy output.

MASS   A given quantity of matter of any kind.

METEOR   The hot, gaseous remains, seen as a bright streak in the atmosphere, resulting from a meteoroid entering the atmosphere at high speed and vaporizing due to frictional heating.

METEORITE   Any meteoroid of stone or metal that survives the journey through Earth's atmosphere and strikes the surface.

METEOROID   Any one of various rock and/or metal fragments orbiting the Sun singly or in swarms.

MOLECULE   Two or more atoms chemically combined. The smallest piece of a substance above the size of an individual atom of that substance. The smallest quantity of water, for example, is a water molecule consisting of 1 atom of oxygen and 2 atoms of hydrogen. One liter of oxygen contains 25 $\times$ 10^{21} molecules. Although individual atoms are too small to be seen, even with the most powerful microscopes, certain giant organic molecules can be seen with electron microscopes.

NEBULA A great cloud of dust and gas within a galaxy. Some nebulae said to be reflection nebulae, reflect light generated by nearby stars, or by stars embedded within the nebula. Other nebulae are dark and so are called dark nebulae. Still others reradiate energy emitted by stars embedded in the nebulae and are called emission nebulae. And still others take the form of a great shell of gas cast off by an eruptive or explosive, star. These are called planetary nebulae because they were once mistaken for planets within the Solar System.

NEUTRON An electrically neutral particle in the nucleus of all atoms except those of hydrogen. Neutrons are only slightly more massive than protons. Outside the atom, neutrons have a life of only 20 minutes or so before decomposing into an electron and a proton and giving off gamma rays.

NEUTRON STAR A star made up of neutrons. Because neutrons are without an electrical charge, they can be packed very closely together. Consequently, neutron stars are extremely dense objects.

NOVA A star that for some reason not yet fully understood bursts into brilliance. Within a few days a typical nova may become hundreds of thousands of times brighter than usual, then it becomes somewhat less brilliant, and after a few months or longer the star returns to its pre-nova brightness. Certain planetary nebulae may be the result of nova eruptions. (See also Supernova.)

NUCLEAR FUSION The union of atomic nuclei and as a result the building of the nuclei of more massive atoms.

Hydrogen nuclei in the core of the Sun fuse and build up the nuclei of helium atoms. In the process large amounts of energy are emitted, thus accounting for the Sun's energy output.

NUCLEUS   In astronomy, the central portion of a galaxy. In chemistry and physics, the central portion of an atom.

NUMBER DENSITY   The number of objects in a given volume. The number density of a classroom of 20 students would be 20 students per room volume. Unlike density, number density is not concerned with mass.

PARALLAX   The apparent change in position of an object when it is viewed from two different positions. The object appears to change place against the background of more distant objects. When you look at a nearby tree first through one eye and then through the other eye by winking rapidly with first one eye and then the other, the tree appears to jump back and forth against the background. By viewing a nearby star from opposite points on Earth's orbit, astronomers can measure the star's angle of apparent shift, the parallax, and so determine the star's distance.

PERIOD   The time a variable star takes to complete one cycle of going from bright to dim and back to bright again. The periods of some variables are measured in hours, while the periods of others are measured in weeks or months. Also, the length of time it takes one celestial object to complete one orbit about another.

PERIOD-LUMINOSITY LAW A relationship between the characteristic periods and mean luminosities of certain variable stars, such as the RR Lyrae and Cepheid variables, that enables astronomers to use these stars as celestial metersticks. The RR Lyrae variables, for example, provided astronomers with a distance scale for our home galaxy. The Cepheid variables have enabled astronomers to determine the distance to certain other galaxies.

PI (π) The sixteenth letter of the Greek alphabet and a symbol designating the ratio of the circumference of a circle to its diameter. The value of pi to eight decimal places is 3.14159265.

PLANET Celestial objects that shine by reflected light from a star about which they are held gravitationally captive and revolve. There are nine known primary planets in the Solar System.

PLANETARY SYSTEM Any star accompanied by one or more planets. Our home Solar System is presumably but one of numerous planetary systems in the Galaxy.

POPULATION I STAR A star, like the Sun, which is associated with interstellar dust and gas in the spiral arms of the Galaxy. Population I stars are relatively young stars formed at least partly by material cast off by aging stars.

POPULATION II STAR A star making up the dense nucleus of the Galaxy and contained in the globular clusters. Population II stars are not associated with interstellar dust, and so are not believed to be forming now. They are old stars.

There seem to be about 10 times as many Population II stars in the Galaxy as there are Population I stars.

PRESSURE   A measurement of force per unit area.

PROMINENCES   Glowing masses of gas that loop and surge from the Sun's surface gases for distances of hundreds of thousands of kilometers. They can best be seen during an eclipse of the Sun or by special instruments such as the coronagraph and the spectrohelioscope.

PROTON   A fundamental particle present in the nucleus of all atoms. A proton has a positive charge of electricity equal in strength to the negative charge of an electron, but a proton is 1,840 times more massive than an electron.

PROTOSTAR   A newly forming star that has not yet begun to radiate visible energy as a result of fusing hydrogen nuclei into helium nuclei.

PULSAR   A rapidly rotating neutron star that sends out a beam of radiation. When the beam points in the direction of Earth, we receive a pulse; about 150 pulsars have been detected, and each has its own rate of pulsation.

PULSATING VARIABLE   A class of variable star that pulses by brightening, dimming, and brightening again in regularly occurring periods.

RADIOACTIVITY   A property of a few of the 100 or more known elements, in which, as they "decay" naturally, and in the process change into a different element, they emit some of the charged particles (protons) in their nuclei.

RED GIANT An enormous star that shines with a red light because of its relatively low surface temperature (about 3,000 K). It is now thought that most, if not all, stars go through a red giant stage after they exhaust their core hydrogen and the core collapses gravitationally. The star then swells up, becoming a red giant.

REVOLUTION The motion of one body around another. The Moon revolves about Earth; the planets revolve about the Sun.

ROTATION The motion of a body around its axis. The Sun and all of the planets rotate, with Earth completing one rotation about every 24 hours.

RR LYRAE VARIABLE A pulsating variable star with a period of less than one day. Because the periods of these stars are closely associated with their average luminosities, the stars can be used as distance indicators.

SCATTERING OF LIGHT The dispersal of light as it passes through any medium, by tiny particles making up the medium. Blue light is scattered by the atmosphere more than light of other wavelengths, and that is why the sky appears blue. Scattering occurs when particles making up the medium are small compared with the wavelength of the light.

SHELL STAR A star that casts off matter explosively and so gives rise to a thin shell of gas that expands around the star.

SOLAR CONSTANT The amount of energy a given area of a planet receives from the Sun per unit time. The solar constant for Earth is about 2 cal/cm²/min.

SOLAR SYSTEM The Sun, its 9 known primary planets accompanied by about 3 dozen known satellites, plus many lesser objects, including comets, the asteroids, meteoroids, and one planetoid.

SPECTROSCOPE 1. Prism spectroscope—an instrument fitted with a prism that separates a star's light into its individual colors, or spectrum. 2. Diffraction grating spectroscope—a spectroscope fitted not with a prism but with a grooved, polished glass surface with many thousands of grooves per centimeter. When a star's light falls on this grating, the different colors making up the light are diffracted at different angles and so form a spectrum. A diffraction grating spectrum of a star is sharper than a prism spectrum.

STAR A hot, glowing globe of gas that emits energy. The Sun is a typical, and our closest, star. Most stars are enormous compared with planets, containing enough matter to make thousands of Earth-like planets. Stars generate energy by the fusion of atomic nuclei in their dense and hot cores. Stars seem to be formed out of dense clouds of gas and dust, evolve through various stages, and finally end their "lives" as dark, cold objects called black dwarfs.

SUPERNOVA A giant star whose brightness is tremendously increased by a catastrophic explosion. Supernova stars are many thousands of times brighter than ordinary nova stars. In a single second, a supernova releases as much energy as the Sun does over a period of about 60 years.

TEMPERATURE   A measure of how hot or cold a body is, "hotness" meaning the rate of atomic motion, or kinetic energy. The greater the kinetic energy, the "hotter" a substance is said to be.

VARIABLE STAR   A star whose energy output varies, either regularly or irregularly. (See also Cepheid Variable, Explosive Variable, Nova, Pulsating Variable, and Supernova.)

VOLUME   A given amount of space.

WAVELENGTH   The distance between two successive crests or troughs of a wave of any kind. The wavelength is found by dividing the velocity of a wave by its frequency.

WHITE DWARF   A very small star that radiates stored energy rather than new energy generated through nuclear fusions. The Sun is destined to become a white dwarf after it evolves through the red giant stage.

ZODIAC   The zone 16° wide stretching around the sky and centered on the ecliptic. The Moon, Sun, and most of the planets revolve about the Sun in this zone, as seen from Earth.

APPENDIX

The Metric System

The universal language of measurement among scientists is the *metric system*. It is a much easier system to learn and use than our English system of measurement.

The basic unit of length in the metric system is the *meter* (about one yard). All other units in the system are related to the meter. Each unit smaller than the meter is one-tenth of the next larger unit, and each unit larger than the meter is 10 times the next smaller unit. Such a system is called a *decimal system*.

UNITS OF LENGTH IN THE METRIC SYSTEM

10 millimeters (mm)	=	1 centimeter (cm)
10 centimeters	=	1 decimeter (dm)
10 decimeters	=	1 meter (m)
10 meters	=	1 dekameter (dkm)
10 dekameters	=	1 hectometer (hm)
10 hectometers	=	1 kilometer (km)

When the metric system was first set up, the basic unit of mass was called the *gram*. One gram is the mass of the amount of water that would fill a cube one centimeter along each edge. One thousand grams are equal to one kilogram (about two pounds). The amount of space that one kilogram of water takes up—1,000 cubic centimeters—is called one *liter* (about one quart). So units for measuring mass and volume in the metric system are based on the basic standard of length, the meter.

UNITS OF MASS IN THE METRIC SYSTEM

10 milligrams (mg)	=	1 centigram (cg)
10 centigrams	=	1 decigram (dg)
10 decigrams	=	1 gram (g)
10 grams	=	1 dekagram (dkg)
10 dekagrams	=	1 hectogram (hg)
10 hectograms	=	1 kilogram (kg)
1,000 kilograms	=	1 metric ton

UNITS OF VOLUME IN THE METRIC SYSTEM

10 milliliters (ml)	=	1 centiliter (cl)
10 centiliters	=	1 deciliter (dl)
10 deciliters	=	1 liter (l)
10 liters	=	1 dekaliter (dkl)
10 dekaliters	=	1 hectoliter (hl)
10 hectoliters	=	1 kiloliter (kl)

CONVERSIONS FOR LENGTH

1 millimeter	=	0.039,37 inch
1 centimeter	=	0.39 inch
1 decimeter	=	3.94 inches
1 dekameter	=	32.81 feet
1 kilometer	=	0.62 mile
1 inch	=	2.54 centimeters
1 foot	=	0.31 meters
1 yard	=	0.91 meter
1 mile	=	1.61 kilometers

CONVERSIONS FOR AREA

1 square millimeter	=	0.002 square inch
1 square centimeter	=	0.16 square inch
1 square decimeter	=	15.50 square inches
1 square kilometer	=	0.39 square mile
1 square inch	=	6.45 square centimeters
1 square foot	=	929.03 square centimeters
1 square yard	=	0.84 square meter
1 square mile	=	259.00 hectares

CONVERSIONS FOR VOLUME

1 cubic centimeter	=	0.06 cubic inch
1 cubic decimeter	=	61.02 cubic inches
1 cubic inch	=	16.39 cubic centimeters
1 cubic foot	=	28.32 cubic decimeters
1 cubic yard	=	0.77 cubic meter

CONVERSIONS FOR MASS

1 milligram	=	0.02 grain
1 kilogram	=	2.21 pounds
1 metric ton	=	2,204.62 pounds (or 0.98 gross ton)
1 ounce	=	31.10 grams
1 pound	=	453.59 grams
1 ton	=	2,240.00 pounds (or 1.02 metric tons)

An Easy Way to Handle Big Numbers

You have found that the Sun has been pouring out energy at its present rate for at least 400,000,000 years. Astronomers often have to work with numbers much larger than that, and it would be a nuisance for them forever to be writing numerous zeros after numbers. One reason it would be a nuisance is that it is easy to lose track of the correct number of zeros in a long string: for instance, when we say that the mass of the Sun is 2,000,000,000,000,000,000,000,000,000,000,000 grams.

To keep track of all those zeros, astronomers use a shorthand system of writing numbers, called *exponential notation*, which uses powers of 10. Here is how it works:

The first power of 10 is 1×10, or 10

The second power of 10 is $1 \times 10 \times 10$, or 100

The third power of 10 is $1 \times 10 \times 10 \times 10$, or 1,000

What is the fourth power of 10?

To write a power of 10 we use an "exponent." For example, we write the fourth power of 10 by writing the exponent 4 up by the right-hand edge of the 10, like this: 10^4. And we read it as "ten to the fourth." So 10^4 is simply 1 with 4 zeros after it, or 10,000. It's that simple. And 10^{10} (ten to the tenth) is 1 with 10 zeros after it, or 10,000,000,000.

Now suppose that you wanted to write 47,000 the shorthand way. You would write it 47×10^3, or 47 with 3 zeros after it. This must be correct because $10^3 = 1,000$, and $47 \times 1,000 = 47,000$. Again, the shorthand way of writing the Sun's mass is 2×10^{33} grams, or 2 with 33 zeros strung out after it.

Now suppose that you wanted to multiply two large numbers expressed in powers of 10. For example,

$$(2 \times 10^{20}) \times (2 \times 10^{40})$$

The first thing we do is multiply 2 × 2, which gives us 4. Then all we do is add the exponents, 20 + 40 = 60. So our answer is 4 × 10^{60}. The thing to remember here is that when you multiply numbers expressed in exponential notation you always add the exponents.

Suppose that we want to divide two numbers expressed in powers of 10. For example,

$$\frac{4 \times 10^{40}}{2 \times 10^{20}}$$

When we divide numbers expressed in exponential notation we subtract the exponents. So the answer to this problem is 2 × 10^{20} (the 2 resulting from 4 ÷ 2 = 2; and the 10^{20} resulting from $10^{40} - 10^{20}$).

Atomic Clocks

Because the "atomic clocks" used to measure the age of rocks, fossils, and other materials play such an important role in science, it is important to understand how these clocks work.

Since the late 1800s, scientists have known that the atoms of certain chemical elements said to be radioactive—such as uranium, potassium, and thorium—break down naturally. They "break down" by giving up bits and pieces of themselves. In the process they turn into the atoms of other elements. For example, uranium turns into lead, and so does thorium. Potassium turns into the element argon. It is easy to understand how these atomic clocks work if you imagine a large box containing 12,800 black marbles representing a radioactive element (Diagram 1). You want to find out how old the box of marbles is. Imagine that the black marbles age (break down) by turning gray. In one year, half of the black marbles turn gray (Diagram 2). At the end of the first year, there would be 6,400 black marbles left, plus 6,400 gray ones. At the end of the second year, half of the remaining 6,400 black marbles turn gray (Diagram 3). There would then be 3,200 black and 9,600 gray marbles. This process would go on and on until our clock ran down.

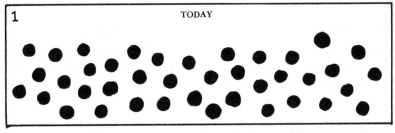

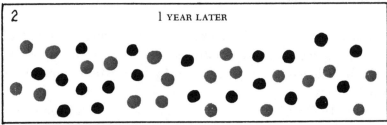

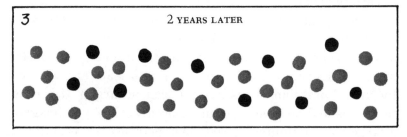

Passage of Time in Years	Number of Black Marbles	Number of Gray Marbles	Black-Gray Ratio
0	12,800	0	0
1	6,400	6,400	1 : 1
2	3,200	9,600	1 : 3
3	1,600	11,200	1 : 7
4	800	12,000	1 : 15
5	400	12,400	1 : 31
6	200	12,600	1 : 63
7	100	12,700	1 : 127
8	50	12,750	1 :255
9	25	12,775	1 : 511

If you counted 800 black marbles and 12,000 gray ones, you would know that four years had gone by since the first black marble turned gray. But can you think of an easier way to tell how much time went by without counting *all* the marbles? Simply scoop a jarful of marbles—maybe 1,000— out of the box, count how many black and gray ones you have in the sample, and divide the number of gray ones by the number of black ones. This is the black-to-gray ratio (see right-hand column of the table). In your sample of 1,000 marbles, there would be 61 black ones and 939 gray ones if the box were four years old, giving a ratio of 1 :15.

That is how a radioactive clock works. Scientists measure the ratio between the number of unchanged atoms of a radioactive element and the number of new atoms that have been formed. The amount of time needed for half the atoms

of a radioactive element to change is called its *half-life.*
Nothing seems to affect the half-life of any radioactive element—neither changes in temperature nor changes in pressure. Since scientists know the half-life of the radioactive
element and since they can measure the ratio of the numbers of new and old atoms, they can then tell how long the
clock has been running since the substance solidified. It is
important that no decay products have been lost since the
substance began decaying. Different radioactive elements
have different half-lives. In the table below are four radioactive elements that are used to date materials, three to
date rocks, and one (carbon) to date material that was once
alive.

THIS RADIOACTIVE ELEMENT	CHANGES INTO	AND HAS A HALF-LIFE OF
uranium-238	lead-206	4,510 million years
potassium-40	argon-40	1,350 million years
rubidium-40	strontium-87	6 million years
carbon-14	nitrogen-14	5,730 years

Usually, only very small amounts of a radioactive material are present in a rock being dated. This means that the
slightest error in measurement may mean a large error in
the ratio between the two elements. It would be like taking a
sample of only 10 marbles, rather than 1,000. The point is
that a sample of 1,000 will be more representative of the
total of the blacks and grays among 12,800 marbles than
would a sample of only 10 marbles. A slight error in

measurement of a small radioactive sample may mean a difference of millions of years in the final absolute age figure. Just a 5 percent error in a 100-million-year old rock might mean an error of 5 million years.

Scientists make two assumptions when they date materials by the radioactive decay method. First, in the potassium-argon series, for example, they assume that no potassium or argon atoms have been added to the rock in question since it was formed and, second, that no argon atoms were present in the rock originally. Although many rocks and minerals are known to have contained some of the decay product (argon in our example), geologists can determine how much of that decay product there was originally. They can then subtract that amount from the total amount of decay product measured and so end up with an accurate measurement of the amount produced by radioactive decay.

INDEX